CLIMATE CHANGE IN THE ANTHROPOCENE

CLIMATE CHANGE IN THE ANTHROPOCENE

KIERAN D. O'HARA

University of Kentucky, Lexington, KY, United States

ELSEVIER

Elsevier
Radarweg 29, PO Box 211, 1000 AE Amsterdam, Netherlands
The Boulevard, Langford Lane, Kidlington, Oxford OX5 1GB, United Kingdom
50 Hampshire Street, 5th Floor, Cambridge, MA 02139, United States

Notices
Knowledge and best practice in this field are constantly changing. As new research and
experience broaden our understanding, changes in research methods, professional
practices, or medical treatment may become necessary.

Practitioners and researchers must always rely on their own experience and knowledge in
evaluating and using any information, methods, compounds, or experiments described
herein. In using such information or methods they should be mindful of their own safety
and the safety of others, including parties for whom they have a professional responsibility.

To the fullest extent of the law, neither the Publisher nor the authors, contributors, or
editors, assume any liability for any injury and/or damage to persons or property as a
matter of products liability, negligence or otherwise, or from any use or operation of any
methods, products, instructions, or ideas contained in the material herein.

ISBN: 978-0-12-820308-8

For Information on all Elsevier publications visit our website at
https://www.elsevier.com/books-and-journals

Publisher: Candice Janco
Acquisitions Editor: Marisa LaFleur
Editorial Project Manager: Jose Paolo Valeroso
Production Project Manager: Bharatwaj Varatharajan
Cover Designer: Christian J. Bilbow

Working together
to grow libraries in
developing countries

www.elsevier.com • www.bookaid.org

Typeset by Aptara, New Delhi, India

CONTENTS

Preface   *ix*

PART I   **1**

1   Our globally changing climate   **3**

   1.1 Introduction   3
   1.2 Global temperature   4
   1.3 Land surface temperature   5
   1.4 Sea surface temperature   6
   1.5 Global surface temperature   7
   1.6 Trends in global temperatures   7
   1.7 Trends in global precipitation   8
   1.8 Extreme weather events   9
   1.9 Changes in the cryosphere   10
   1.10 Changes in sea level   14
   1.11 Changes in land processes   14
   References   16

2   Physical drivers of climate change   **19**

   2.1 The global radiation budget   19
   2.2 The greenhouse effect   19
   2.3 Radiation forcing   22
   2.4 Global warming potential   24
   2.5 Greenhouse gases   25
   2.6 Aerosols   29
   2.7 Climate response   30
   2.8 Feedbacks   32
   2.9 Albedo feedbacks   34
   2.10 Ocean chemistry, ecosystems, and circulation   35
   2.11 Permafrost   38
   References   38

3   Evaluation of climate model performance   **41**

   3.1 Introduction   41
   3.2 Model types   42
   3.3 Model improvements   43

3.4 Model evaluation   44

3.5 Ensemble approach to evaluation   45

3.6 Model intercomparisons   45

3.7 Results   46

3.8 The ocean   48

3.9 Carbon cycle   52

3.10 The Paris Accords   53

3.11 Representative climate pathways   53

3.12 Near-term climate projections   54

3.13 Long-term projections   57

References   60

4 Paleoclimates   **63**

4.1 Introduction   63

4.2 Preindustrial external radiative forcings   65

4.3 High CO_2 worlds   67

4.4 Pleistocene glacial-interglacial dynamics   69

4.5 The CLIMAP Project   71

4.6 Holocene climate   73

References   76

PART II   **79**

5 Climate impacts: US sectors and regions   **81**

5.1 Introduction   81

5.2 Key sectors   81

5.3 Regional climate impacts   95

References   101

6 Adaptation   **105**

6.1 Introduction   105

References   121

7 Mitigation   **123**

7.1 Introduction   123

7.2 GHG emission trends   125

7.3 Emission drivers   127

7.4 Carbon intensity of energy   129

7.5 Sectors   130

7.6 Buildings   136

7.7 Shared socioeconomic pathways – quantifying the paths   137
7.8 Comparison of SSP1 and SSP3   138
7.9 SSP5. Fossil fuel development   139
References   140

PART III   **143**

8   1.5°C versus 2.0°C warming   **145**
8.1 Introduction   145
8.2 1.5°C and 2.0°C warming   147
8.3 Natural systems   149
8.4 Human systems   153
References   155

9   Getting to net zero by 2050   **157**
9.1 Introduction   157
9.2 The current situation (2021)   158
9.3 Road to net-zero emissions 2050   160
9.4 Population and GDP   161
9.5 Energy and CO_2 prices   161
9.6 CO_2 emissions   162
9.7 Total energy supply   162
9.8 Economic sectors   163
9.9 Conclusions   164
References   164

10 Climate engineering   **167**
10.1 Introduction   167
10.2 Solar radiation management   168
10.3 Aerosol injection into the stratosphere   169
10.4 Albedo enhancement of low-level marine clouds   174
10.5 Surface albedo enhancement   176
10.6 Carbon dioxide removal   176
10.7 Discussion   183
References   184

Index   *187*

Preface

The Greek word for human kind is anthropos. The term Anthropocene was proposed over two decades ago by Paul Crutzen (atmospheric scientist and Nobel laureate) and Eugene Stoermer (biologist) to indicate a new geological epoch in which the intensity of human activity strongly impacted Earth Systems, thereby marking the end of the current Holocene epoch, and justifying a new epoch. The Anthropocene has not been formalized as a new geologic epoch and even the boundary between it and the earlier Holocene has not yet been agreed upon, but the term nevertheless has gained widespread currency in both the scientific and popular literature.

This book follows the original suggestion that the Industrial Revolution marks the beginning of the Anthropocene, marked by the transition from a pastoral lifestyle to an industrial one largely based in cities (circa 1800 AD). This time frame corresponds to an increase in burning of coal and increased emissions of greenhouse gases, especially carbon dioxide. Based on ice cores, the preindustrial atmospheric concentration of carbon dioxide was about 280 ppm (compared to $\sim$420 ppm in 2020) and is commonly used as a reference point when discussing climate change. By 2017, the global mean surface temperature had increased by 1.0°C ($\pm$ 0.2) (1.8°F) since preindustrial times, and both of these reference frames are used throughout the book.

The concept of the Anthropocene provides a lens through which insight into man's effects on the environment can be viewed in a structured historical fashion. It is worth noting that the geological community on altering the geological time scale moves at a glacial pace: in 1878, Charles Lapworth, proposed the Ordovician Period to be placed between the younger Silurian Period and the older Cambrian Period; the proposal was formally accepted in 1976.

This book is to a large extent based on the Intergovernmental Panel on Climate Change (IPCC) reports. The World Meteorological Organization (WMO) together with the United Nations provides the basis for these reports which are published approximately every five or six years. The United States Government's Fourth National Climate Assessment (NCA4, 2017), with input from 13 government agencies, is also heavily relied upon and its conclusions agree closely with those of the IPCC reports. The fifth IPCC report (IPCC-AR5) was published in 2013–2014 and the latest report

(IPCC-AR6) was published in August of 2021, having been delayed by the global pandemic of 2020. Report volumes are divided into three working groups (WG1, II, III), and each chapter commonly has twenty or more international expert authors and each volume is weighty, often at a thousand pages or more per volume. The peer review process of these reports has several rounds and is extensive and lengthy. This book is largely a summary of these reports.

Following Caesar's Gaul, the book is divided into three parts. Part I addresses the physical science basis of climate change and is largely based on IPCC-AR5 (2013). Chapter 1 addresses the basic observations indicating climate change, followed by the drivers of this change in chapter 2. Chapter 3 examines computer climate models and chapter four looks at paleoclimate reconstructions. Part II examines climate impacts in various regions of the USA (based on NCA4, 2017), followed by adaptation and mitigation scenarios. Part III looks at the difference between 1.5 and 2.0°C warming risks (based on IPCC Special Report, 2018) followed by the road map to net–zero emissions by 2050 (based on the International Energy Agency 2021 report). The final chapter examines climate engineering (or geoengineering), which is widely regarded as a last resort option, and this chapter is based on the current scientific literature.

Although Anthropos applies to all humanity, it is clear that, based on geography and socioeconomic status, the impacts of climate change are related to social inequities and the impacts are not and will not be distributed evenly– the developing countries and the poor will be most affected. The Paris Agreement of 2015 recognized this fact but whether the developed countries will fulfill their monetary promises to developing nations remains in doubt. The United States re-entered the Paris agreement in 2020. The United Nations climate summit of November 2021 (COP 26), held in Glasgow, agreed to reduce methane emissions (by 30%) by 2030 and also to eliminate deforestation by the same date. No agreement to a coal ban was reached, as China, India and Russia did not sign on.

PART I

1. Our globally changing climate   3
2. Physical drivers of climate change   19
3. Evaluation of climate model performance   41
4. Paleoclimates   63

CHAPTER 1

Our globally changing climate

1.1 Introduction

The Earth sciences study a multitude of processes that shape the spatial and temporal character of our environment (Fig. 1.1). Modern day observations, archives of past climates, climate model projections, and statistical tools, can all be used to yield significant insight into climate change, resulting in conclusions that have variable levels of confidence from high to low (see Cubasch et al., 2013). The Earth's climate system is powered by solar radiation about half of which is in the visible and ultraviolet range of the electromagnetic spectrum. The sun provides its energy primarily to the tropics, which is redistributed to higher latitudes by atmosphere and ocean transport processes. The relatively cool temperature of the Earth's surface means it reradiates energy in the long wavelength part of the spectrum (infrared) and much of this radiation is absorbed by gases in the atmosphere such as water vapor, CO_2, CH_4, and N_2O as well as halocarbons – this is the greenhouse effect. Given the Earth has had a near constant temperature over the past few centuries the incoming solar energy must nearly balance the outgoing energy to space, and clouds play an important role in this energy balance. About 30% of the shortwave radiation is reflected back to space by clouds, causing cooling. On the other hand, some clouds, depending on elevation, trap long wave radiation, heating the surface, and the lower atmosphere.

Climate is average weather over a prolonged period, commonly taken as three decades or longer, and climate change refers to a change in the state of the climate (based on statistical tests), such as temperature, precipitation, or drought. For example, during the last glaciation, stadial, and interstadial periods were characterized by cold/dry climates (stadials) alternating with warm/wet climates (interstadials), on a millennial time scale (O'Hara, 2014). Fig. 1.1 summarizes several key elements of the climate system; elements interact with one another in complex ways involving both positive and negative feedbacks (see Chapter 2). This chapter summarizes several indicators that our planet is currently warming. The warming dates back to the beginning of the anthropocene, where the mean temperature over the period 1850–1900 is taken as the reference period.

Climate Change in the Anthropocene.
DOI: https://doi.org/10.1016/B978-0-12-820308-8.00005-2

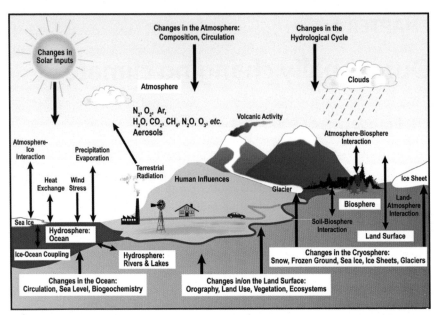

Figure 1.1 Summary of major drivers of climate change. *(Source with permission: Cubasch et al., 2007.)*

1.2 Global temperature

The fourth IPCC assessment report (LeTreut et al., 2007) provides a history of early attempts at constructing a global temperature time series for the nineteenth and twentieth centuries. The global average temperature is one of the most important variables in the study of climate change as it correlates with other variables such as ice melting, sea level rise, precipitation, and because it has the most robust record over time. The concept of a global average temperature is simple in principle but its calculation is far from trivial (Vose et al., 2012). Although the thermometer was invented as early as the 1600s it was not until the 1900s that different global estimates of average land temperature began to agree with one other.

The German climatologist W. Köppen (1846–1940) was one of the first to recognize the major problems involved in the global average temperature estimates namely, access to data in usable form, quality control to remove erroneous data, standardization to ensure fidelity of data, and area averaging in areas of substantial data gaps. Köppen averaged annual observations from 100 stations into latitude belts to produce a near global time series as early as the late nineteenth century. The International Meteorological Organization (IMO) formed in 1873, and its successor the World Meteorological Organization (WMO), still work to promote and standardize observations. The

World Weather Records (WWR), formed by the IMO in 1923, provided monthly data for temperature (and also pressure and precipitation) estimates from hundreds of stations in the early twentieth century with data beginning in the early 1800s. Callendar (1938) used these data to provide one of the first modern land-based global average temperature time series. As mentioned in the Preface, the World Meteorological Organization (WMO) together with the United Nations today provides the basis for the IPCC scientific reports on climate change and on which this book is largely based.

Today, three research groups study global sea and land-based temperatures put together from piecemeal records (Vose et al., 2012): the National Oceanic and Atmospheric Administration's National Climatic Data Center (NOAA-NCDC), the National Aeronautic and Space Administration's Goddard Institute for Space Studies (NASA-GISS) and the Met Office Hadley Center and Climatic Research Unit (HadCRUT). Each group uses somewhat different input datasets and they also analyze the data with different methodologies. For example, GISS makes extensive use of satellite data, whereas NCDC uses it in a limited capacity and HasCRUT makes no use of satellite data. Similarly, GISS and NCDC provide temperature estimates in unsampled areas (using interpolation), whereas HasCRUT does not. Despite these differences all three groups reach a similar conclusion: since 1900 the global average surface temperature increase has been about $0.8 \pm 0.2°C$. The fifth Intergovernmental Panel on Climate Change (IPCC-AR5, 2013) and the US government's Fourth National Climate Assessment (NCA4, 2017) reports both agree with this conclusion with a high level of confidence. These reports also project that by the end of this century (2100) the global average temperature increase will be between 2.0°C and 5.0°C, depending on greenhouse gas emissions and population and economic growth among other variables (see Chapter 6).

1.3 Land surface temperature

The dataset used by NCDC consists of historical monthly data going back a century from over 7000 surface weather stations. The data set is reviewed for quality assurance and spatial inconsistencies. Land surface temperatures require adjustments due to a variety of causes such as station relocation, change in instrumentation (e.g., automation), urbanization (the city heat effect) and land use, and microclimate changes. Such changes typically produce an abrupt jump relative to its neighbor stations. These artifacts are indentified automatically by comparing surrounding stations pair wise. Reno Nevada, for example, required an adjustment of 2°C after the station was moved from down town to the airport (Thorne, 2016). The transition to electronic sensors in the US in the late twentieth century required an adjustment of about 0.25°C nationwide (Vose et al., 2012). Averaged over the globe, however, these adjustments have only a minor impact on the long-term LST record.

The temperature series is also standardized to account for elevation, latitude, coastal proximity, and season. A mean temperature is calculated for each station relative to the reference period (1961–1990) and then this mean is subtracted from each temperature value at that station. The resulting values are referred to as anomalies and this is the most common way the results are presented in graphic form. This standardization procedure reduces much of the variability in the original dataset.

The uneven spatial distribution of stations is taken into account by averaging measurements in 5-degree longitude and latitude grid boxes. A single average temperature is calculated for each box on a monthly and annual basis and this helps prevent high-density measurement boxes to have undue influence. Today land coverage is about 90% and areas of low coverage include forests, deserts and the poles. Satellite data affords global coverage but the data must be calibrated with ground measurements; in addition, because an infrared spectrometer is used for temperature measurements, the skies must be cloud-free.

1.4 Sea surface temperature

The sea surface temperature dataset is primarily from marine meteorological observations from buoys and ships integrated from numerous historical sources. Buoys can be either drifting or moored; buoy observations are given about six times the weight from ships on account of the noise in the latter observations (e.g., mistakes in navigation, instrument calibration, data transcription). Ship temperature measurements show a change in practice over time. In pre-World War II times wooden or canvas buckets (some insulated, some not) were hauled on deck for measurement. These measurements require adjustments for several variables: type of bucket, height of deck, etc. Evaporative cooling, especially in high winds, requires adjustments of about 0.2°C (Thorne, 2016). Later on, the measurements were made at the engine's cool water intake, or sensors were placed on the ship's hull. Globally, a smaller grid box (compared to the LSTs) of 2 × 2 degrees is used. Each box value is an average of measurements over a month and the mean value for a reference time period (197 –1990) is subtracted from each temperature measurement, as in the case for LSTs.

1.5 Global surface temperature

Before merging the LST and SST anomalies they are processed separately because there are fundamental differences between the two datasets (Vose et al., 2012). First, the spatial coverage over the oceans is substantially less than that over land (Thorne, 2016) and secondly, the density of ocean measurements is substantially lower than land measurements. In addition, the time and space scales of temperature variability over land are shorter compared to the ocean, due to the higher specific heat of water and its slower speed of advection. Before merging the datasets, low frequency variations that occur over longer periods and high frequency variations that occur over shorter periods are identified and smoothed, then both components are added together. The LST and SST datasets are merged after the SST grid boxes (2° x 2°) are averaged into 5° x 5° boxes. The reference time period over the ocean (1971–2000) is converted to the same time period as the land measurements (1961–1990). Other adjustments are described in more detail in Vose et al. (2012). The global yearly and monthly averages are simply the average of all boxes having a value in that year and month. The annual global average temperature is simply the arithmetic mean of 12 monthly averages.

1.6 Trends in global temperatures

Fig. 1.2 shows the NCA4 annual (top) and decadal (bottom) average global temperatures over land and ocean for the period 1880–2016, relative to the reference period 1901–1960 (Wuebbles et al., 2017). The global annual average temperature has increased by 0.7°C (1.2°F) for the period 1986–2016. Year to year fluctuations are due to natural variations such as El Niños and La Niñas and volcanic eruptions. On a decadal scale (bottom diagram) these fluctuations are smoothed out and every decade since 1966-1975 has been warmer than the previous decade. Recent decades show greater warming due to accelerating greenhouse gas emissions. Sixteen of the 17 warmest years since the late 1800s occurred in the period from 2001 to 2016. In general, winter is warming faster than summer and nights are warming faster than days.

Fig. 1.3 shows the global surface temperature (°F) increase for the period 1986–2015 referenced to the 1901–1960 time frame. Note the oceans show less warming compared to the continents on account of their higher heat capacity. On the continents the largest increases are seen in Eurasia, northwest North America, central South America, and northwest Africa.

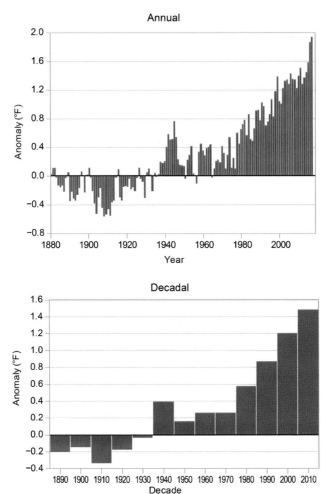

Figure 1.2 Annual (top) and decadal (bottom) combined ocean and land temperatures for the period 1880–2010. *(Source with permission: Wuebbles et al., 2017.)*

1.7 Trends in global precipitation

The Clausius–Clapeyron relation describes the water liquid–vapor equilibrium as a function of pressure and temperature. Global atmospheric water vapor should increase by about 6%/°C to 7%/°C and satellite data over the oceans agree with this estimate (Santer et al., 2007); increases in water vapor should lead to increased precipitation. Global time series of precipitation

Surface Temperature Change

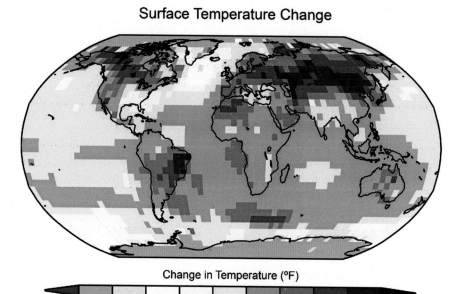

Figure 1.3 Global surface temperature for the period 1986–2015 relative to the 1901–1960 mean. *(Source: NOAA.)*

over the past century show a slight rise but are not statistically significant because of the sparsity of data in the early record (Wuebbles et al., 2017). The global distribution map shows increased precipitation at higher latitudes and lower precipitation at lower latitudes due to Hadley cell circulation. Deficits in precipitation are notable in Africa, the Tibetan plateau and southern China, western USA, and eastern Australia; as expected the Amazon rain forest basin shows higher precipitation.

1.8 Extreme weather events

The distribution of extreme weather events can be approximated by a Normal distribution where extreme events (hot or cold) are rare and correspond to the tails of the distribution (Fig. 1.4). In a warming world the mean of the distribution can be expected to shift to the right giving rise to more extreme hot events and also fewer cold events. Fig. 1.5 shows decreasing number of cold nights and days (top two insets) and increasing number of warms nights and days (bottom two insets) for the period 1950–2010 relative to the reference time period 196–1990. The patterns are similar

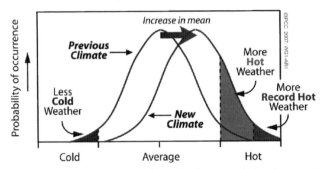

Figure 1.4 Climate extremes occur on the tails of a normal distribution. Cold periods become less common and hot events become more common as the average climate becomes warmer. *(Source with permission: Cubsach et al., 2013.)*

to those predicted by Fig. 1.4. Increases in frequency of extreme precipitation events are expected from an increase in atmospheric water vapor and annual-maximum daily precipitation events have increased 8.5% over the past 110 years over both wet and dry regions (Wuebbles et al., 2017). Computer climate models (Chapter 3) also predict increased extreme precipitation events.

1.9 Changes in the cryosphere

The cryosphere includes continental ice sheets (Greenland and Antarctica), sea ice (e.g., Arctic Ocean), mountain glaciers, frozen ground (permafrost), lake and river ice and snow (Vaughan et al., 2013; Wuebbles et al., 2017). These different components of the cryosphere respond to changing conditions on different time scales: river and lake ice and snow respond on a daily timescale, glaciers on an annual basis, mountain glaciers over centuries, and large ice sheets on a millennial timescale. Changes in land-based parts of the cryosphere have a major impact on sea level change and the areal extent of ice (see Box 1.1). Both on land and sea-based ice have a major influence on sunlight reflectivity or albedo (e.g., open water ~5%; ice ~50–70% and snow covered ice ~90%), which in turn affects climate feedbacks. On a more local scale, melting glaciers may affect tourism and impact freshwater resources. For example, earlier Spring melting of snow in the Sierra Nevada mountains results in increased runoff, depriving downstream aquifers of replenishment and resulting in changing hydroelectric generation schedules. The cryosphere is part of a complex climate system and is one of the best barometers of current climate change. Table 1.1 shows the percent area of

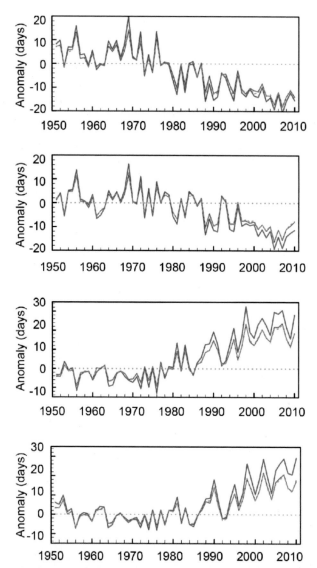

Figure 1.5 Global decreasing cold nights and days (top two insets) and increasing warms nights and days (bottom two insets) over the period 1950–2010. *(Source with permission: Hartmann et al., 2013.)*

various components of the cryosphere on both land and sea. The sea level equivalent if these ice components were to melt is also shown and does not include sea level rise due to thermal expansion of warming oceans. This sea level rise is simply ice volume melt divided by the area of the oceans

Table 1.1 Land and ocean cryosphere components.

Ice on land	Land area (%)	Sea level equivalent (m)
Antarctic ice sheet	8.3	58.3
Greenland ice sheet	1.2	7.4
Glaciers	0.5	0.4
Permafrost	9–12	Not applicable
Seasonally frozen	33	
Lake and river ice	1.1	
Total	53–56	66
Ice in ocean	Ocean area (%)	Volume (10^3 km^3)
Antarctic ice shelf	0.45	380
Antarctic sea ice austral summer (spring)	0.8 (5.2)	3.4 (11.1)
Arctic sea ice autumn (winter)	1.7 (3.9)	13.0 (16.5)
Total	5.3–7.3	

Source with permission: adapted IPCC-AR 5 (2013), table 4.1.

BOX 1.1 Polar amplification

Polar amplification occurs when the magnitude of zonal averaged temperature change at high latitudes exceeds the global average temperature, in response to climate forcings (e.g., orbital forcing or green house gases) on time scales greater than the annual cycle. Amplification has important implications for polar ice sheet stability and hence sea level and also for the carbon cycle involved in permafrost thawing. It has been known since Milankovitch's time that orbital forcing was more important at high latitudes (Imbrie and Imbrie, 1986). Today, orbital forcing is commonly calculated at 50° to 60° latitude in the northern hemisphere (NH). In the Arctic, the sea ice/ocean albedo feedback is also important. Retreating sea-ice decreases the surface albedo causing ocean warming and further melting. In continental Arctic regions snow cover change also changes the surface albedo. Surface vegetation changes occur on a longer timescale (decades to centuries) and also affect the albedo. On glacial-interglacial time scales (thousands of years), the slow retreat of the ice also leads to albedo change and is important to polar amplification in the NH.

In the Southern Ocean, sea surface temperature is amplified in response to changes in radiative forcings, also due to the sea ice/ocean albedo feedback. But as the Southern Ocean is less stratified compared to the Arctic, it can absorb much more heat (a stratified ocean does not allow as much downward heat transport). This means the Southern Ocean temperature response is damped in comparison to the Arctic.

(363×10^6 km^2). Note: an ice melt volume of 363 Gt would cause a 1 mm rise in global sea level; 1Gt (10^9 metric tons) of ice is equivalent to a volume of 1 km^3.

Satellite data from Gravity Recovery and Climate Experiment (GRACE) have provided gravimetric land ice measurements indicating mass loss from the global cryosphere (Velicogna and Wahr, 2006a, 2006b). These measurements indicate mass losses from the Antarctic and Greenland ice sheets and mountain glaciers around the world. The annually averaged mass loss from 37 reference glaciers has increased every year since 1984. Over the period 2003–2009, 19 reference mountain glaciers all show mass loss with Alaska, Greenland, Southern Andes, Canadian Arctic, and the Asian mountains accounting for 80% of global ice loss (Vaughan et al., 2013). Arctic sea ice extent and thickness and volume have all decreased since 1979 when satellite data first began (http://nsidc.org/arcticseaicenews/). Typically, Arctic sea ice is at maximum in March and a minimum in September after summer melting. September sea ice extent decreased by 13.3 percent per decade between 1979 and 2016. Climate models (Chapter 3) project a nearly ice-free Arctic Ocean in summer time by mid-century (2050). The lower albedo of an ice-free Arctic would be substantial and is equivalent to a radiative forcing of 0.3 Wm^{-2} (Chapter 2). Over the past decade mass loss from the Greenland ice sheet has accelerated, losing 244 ± 6 Gt per year between 2003 and 2013. Satellite data for 2012–2013 showed a loss of 562 Gt, twice the annual average. Greenland ice sheet mass loss is by surface melting as well as discharge from the base of the sheet.

West Antarctica is characterized by land ice that transitions to coastal ice and sea ice sheets. Air temperatures are too cold for surface melting in Antarctica even in the summer, and recent ice loss from West Antarctica sea ice is attributed to warming of the surrounding ocean. Evidence suggests that the Amundsen Sea sector is expected to entirely disintegrate by this century, corresponding to a sea level rise of 1.2 meters (Wuebbles et al., 2017). Areas of East Antarctic sea ice show gains of 1.2% to 1.8% since 1979; these gains are much smaller than losses seen in Arctic ice.

Terrestrial permafrost shows a temperature increase of 1°C to 2°C from a variety of northern regions over the period from the 1970s to 2010 in boreholes at a depth of 5 m to 20 m; these depths are below seasonal temperature variations. The southern boundary of permafrost has moved up to 80 km north in some high latitude locations (e.g., Siberia). The thawing of permafrost raises concerns of the release of both CO_2 and CH_4 by bacterial activity resulting in a positive feedback for warming. Other concerns are the impact on infrastructure (instability of roads and buildings). For example, a fuel storage depot in northern Siberia (Norilsk) collapsed in May of 2020 spilling 17,500 metric tons of diesel fuel into the local river that flows to the Arctic Ocean. The collapse was attributed to thawing permafrost.

1.10 Changes in sea level

Statistical analyses of tide gauge data indicate that global mean sea level (GMS) has increased by 20–23 cm since 1880 with a rate of about 1.2 cm to1.5 cm per decade from 1901 to 1990. However, since the early 1990s both tide gauge and satellite altimeters have recorded a faster rate of about 3 cm per decade, resulting in about an 8 cm rise in GMS since the early 1990s. This rise is attributed to two components, two thirds of which is due to melting of land-based ice and one third to thermal expansion of the oceans (Gehrels, 2016).

Future projections of sea level rise depend on which representative climate pathway (RCP) we follow (Cubasch et al., 2013; see also Table 3.1), which in turn depends of future greenhouse emissions, population growth, and societal economic progress among other variables (Chapter 6). Model projections for the high scenario (RCP8.5; similar to what is sometimes termed "business as usual") indicate a sea level rise of 0.5–1.3 m by 2100. This scenario involves a global temperature increase of about 4°C. The low scenario (RCP2.6) indicates a sea level rise of 0.24–0.8 m and this scenario involves a temperature increase of about 2°C by end of the century.

1.11 Changes in land processes

Changes in land cover can have important effects on climate and conversely changes in climate cause changes in land cover leading to feedback mechanisms both positive and negative. Northern hemisphere snow cover has decreased by about 0.5 million km^2 in the Spring due to earlier Spring melting, largely due to global warming since the 1970s, resulting in reduced albedo of the land surface. Globally, land-use change since the 1750s has been typified by deforestation replaced by intensive farming and urbanization, thereby increasing albedo, resulting in a small cooling effect (see Table 2.3 for albedo values). This deforestation has released about 190 ± 65 GtC over this time period. Over the same time period anthropogenic emissions were 600 ± 70 GtC, so that cumulative land use change amounts to about 32% of total emissions. Tropical deforestation by biomass burning emits about 0.1–1.7 GtC per year. Global deforestation emits about 3 GtC per year, but this is

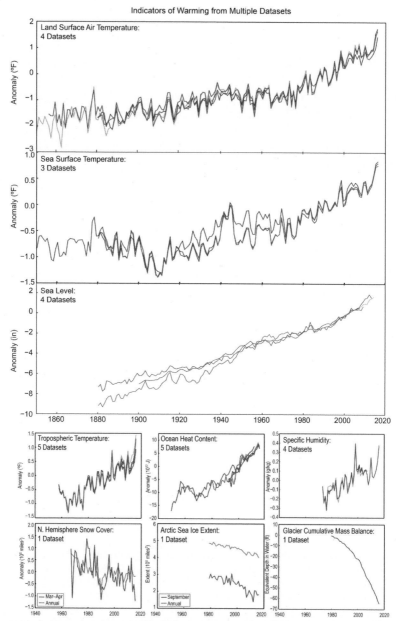

Figure 1.6 Summary diagram of global climate change properties. From top to bottom: land surface temperature rise (4 datasets), sea surface temperature rise (3 datasets), sea level rise (4 datasets); Insets, left to right: Troposphere temperature rise (5 datasets), ocean heat content rise (5 datasets), humidity rise (4 datasets), northern hemisphere snow cover fall (1 dataset), Arctic sea ice fall (annual and September), glacier mass loss (1 dataset). *(Source with permission: Wuebbles et al., 2017.)*

partly offset by regrowth of forest on abandoned agricultural land by about 2 GtC per year. It is only relatively recently that this kind of information has been quantified, mainly due to satellite observations (Wuebbles et al., 2017).

It has been shown that elevated levels of CO_2 lead to increased greening by lengthening the growing season, thereby allowing more carbon storage in live biomass (Wenzel et al., 2016). But this effect is limited by the supply of other nutrients, such as nitrogen and water. Increase in temperature and precipitation in high latitudes increases soil decomposition and emissions of CO_2 and CH_4. It has already been mentioned that permafrost thawing is also likely to be a positive feedback for these greenhouse gases (Fahey et al., 2017). Increased temperature and reduced precipitation increases wildfire risk in terrestrial ecosystems, reducing the potential for carbon storage. Many of these feedbacks have not been quantified and some are not yet fully understood (Chapter 2). Fig. 1.6 summarizes the major indicators of global warming including land surface air temperature, sea surface temperature, sea level, atmosphere temperature, ocean heat content, humidity, northern snow cover, Arctic ice extent and glacier mass.

References

Callendar, G.S., 1938. The artificial production of carbon dioxide and its influence on temperature. Quart. J. Roy. Meteorol. Soc. 87, 1–12.

Cubasch, U., Wuebbles, D., Chen, D., Facchini, M.C., Frame, D., Mahowald, N., Winther, J.-G., 2013. Introduction. In: Stocker, T.F., Qin, D., Plattner, G.K., Tignor, M., Allen, S.K., Boschung, J., Nauels, A., Xia, Y., Bex, V., Midgley, P.M. (Eds.), Climate Change 2013: The Physical Basis. Contribution of Working Group 1 to the Fifth Assessment of the Intergovernmental Panel on Climate Change. Cambridge University Press, UK and New York.

Fahey, D.W., Doherty, S.J., Hibbard, K.A., Romanou, A., Taylor, P.C., 2017. Physical drivers of climate change. In: Wuebbles, D.J., Fahey, D.W., Hibbard, K.A., Dokken, D.J., Stewart, B.C., Maycock, T.K. (Eds.), Climate Science Special Report: Fourth National Climate Assessment, 1. U.S. Global Change Research Program, Washington D. C., USA, pp. 73–113.

Hartmann, D.L., Klien Tank, A.M.G., Rusticucci, M., Alexander, L.V., Brönnimann, Charabi, Y., Dentener, F.J., Dlugokencky, E.J., Easterling, D.R., Soden, B.J., Thorne, B.W., Wild, M., Zhai, P.M., 2013. Observations: atmosphere and surface. In: Stocker, T.F., Qin, D., Plattner, G.K., Tignor, M., Allen, S.K., Boschung, J., Nauels, A., Xia, Y., Bex, V., Midgley, P.M. (Eds.), Climate change 2013: The Physical Basis. Contribution of Working Group 1 to the Fifth Assessment of the Intergovernmental Panel on Climate Change. Cambridge University Press, U. K. and New York.

Gehrels, R., 2016. Rising sea : levels. In: Letcher, T.M. (Ed.), Climate Change. Elsevier, New York.

Imbrie, J., Imbrie, K.P., 1986. Ice Ages – Solving the Mystery. Harvard University Press, Cambridge MA.

IPCC5, 2013. Climate change 2013. The physical science basis. In: Stocker, T.F., Qin, D., Plattner, G.K., Tignor, M., Allen, S.K., Boschung, J., Nauels, A., Xia, Y., Bex, V., Midgley, P.M. (Eds.), Working Group I Contribution to the Fifth Assessment Report of the Intergovernmental Panel on Climate Change. Cambridge University Press, UK and New York.

LeTreut, H., Somerville, R., Cubasch, U., Ding, Y., Mauritzen, C., Mokssit, A., Peterson, T., Prather, M., 2007. Historical overview of climate change. In: Solomon, S., Qin, D., Manning, M., Marquis, M., Averyt, K., Tignor, M.M.B., Miller, H.L., Chen, Z. (Eds.), Climate Change 2007: The Physical Science Basis. Contributions of Working Group I to the fourth Assessment Report of the Intergovernmental Panel on Climate Change. Cambridge University Press, UK and New York.

NCA4, 2017. Climate science special report. In: Wuebbles, D.J., Fahey, D.W., Hibbard, K.A., Dokken, D.J., Stewart, B.C., Maycock, T.K. (Eds.). The United States Government's Fourth National Climate Assessment, 1. U.S. Global Change Research Program, Washington D. C., USA.

O'Hara, K.D., 2014. Cave Art and Climate Change. Archway Publishing, Bloomington, IN.

Santer, B.D., Mears, C., Wentz, F.J., Taylor, K.E., Gleckler, P.J., Wigley, T.M.L., Barnett, T.P., Boyle, J.S., Bruggemann, W., Gillett, N.P., 2007. Identification of human-induced changes in atmospheric moisture content. Proc. Natl. Acad. Sci. 104, 15248–15253.

Thorne, P., 2016. Global surface temperatures. In: Letcher, T.M. (Ed.), Climate Change, Observed Impacts on Planet Earth, 2nd ed. Elsevier, Amsterdam.

Vaughan, D.G., Comiso, J.C., Allison, I., Carrasco, J., Kaser, G., Kwok, R., Mote, P., Murray, T., Paul, F., Ren., J., Rignot, E., Solomina, O., Steffen, .K., Zhang, T., 2013. Observations: Cryosphere. In: Stocker, T.F., Qin, D., Plattner, G.K., Tignor, M., Allen, S.K., Boschung, J., Nauels, A., Xia, Y., Bex, V., Midgley, P.M. (Eds.), Climate Change 2013: The Physical Basis. Contribution of Working Group 1 to the Fifth Assessment of the Intergovernmental Panel on Climate Change. Cambridge University Press, UK and New York.

Velicogna, I., Wahr, J., 2006a. Acceleration of greenland ice mass loss in spring. Nature 443, 329–331.

Velicogna, I., Wahr, J., 2006b. Measurements of time-variable gravity show mass loss in Antarctica. Science 311, 1754–1756.

Vose, R.S., Arnt, D., Banzon, V.F., Easterling, D.R., Gleason, B., Huang, B., Kearns, E., Lawrimore, J.H., Menne, M.J., Peterson, T.C., Reynolds, R.W., Smith, T.M., Willliams Jr., C.N., Wuertz, D.B., 2012. NOAA's merged land-ocean surface temperature analysis. Bull. Amer. Meteorol. Soc. 93, 1677–1685.

Wenzel, S., Cox, P.M., Eyring, V., Friedlingstein, P., 2016. Projeted land photosynthesis constrained by changes in the seasonal cycles of atmospheric CO_2. Nature 538, 499–502.

Wuebbles, D.J., Easterling, D.R., Hayhoe, K., Knutson, T., Kopp, R.E., Kossin, J.P., Kunkel, K.E., LeGrande, A.N., Mears, C., Sweet, W.V., Taylor, P.C., Vose, R.S., Wehner, M.F., 2017. Our globally changing climate. In: Wuebbles, D.J., Fahey, D.W., Hibbard, K.A., Dokken, D.J., Stewart, B.C., Maycock, T.K. (Eds.). Climate Science Special Report. The United States Government's Fourth National Climate Assessment, vol. 1.

CHAPTER 2

Physical drivers of climate change

2.1 The global radiation budget

The global mean radiation budget in the National Climate Assessment Report (Fahey et al., 2017) is similar to that of the earlier IPCC-AR5 report (Hartmann et al., 2013) and is shown in Fig. 2.1. Satellite data indicate the solar irradiance at the top of the atmosphere is 1360 Wm^{-2}. The cross sectional area of the Earth that intercepts this radiation from the sun is πr^2 where r is the radius of the Earth. The surface area of the Earth is $4\pi r^2$ so that the average incoming solar radiation per unit area is 1360 Wm^{-2} divided by four or 340 Wm^{-2} as indicated in Fig. 2.1. Of this amount, 100 Wm^{-2} ($\sim$30%) is reflected back to space at the top of the atmosphere leaving 240 Wm^{-2} absorbed by the Earth (atmosphere and surface) which nearly balances the thermally emitted budget (239 Wm^{-2}) together with the 0.6 Wm^{-2} that is stored within the Earth itself globally.

The Earth surface radiation budget is more uncertain than that for the top of the atmosphere as surface measurements are not globally available and satellites cannot always make good estimates at the surface. It is estimated solar radiation absorbed by the atmosphere is 79 Wm^{-2} and that absorbed at the surface is 161 Wm^{-2} giving a total of 240 Wm^{-2} absorbed by the surface and atmosphere. Satellite and ground measurements indicate a downward flux of infrared radiation, trapped by greenhouse gases of 342 Wm^{-2}. The mean global latent flux due to evaporation is estimated at 84 Wm^{-2}, which should equal precipitation, so that global estimates of precipitation can be used to constrain the magnitude of latent heat of evaporation.

2.2 The greenhouse effect

All bodies emit and absorb radiation at various wavelengths depending on their temperature. The greenhouse effect can be understood in the context of the wavelength emitted and absorbed by the Sun and the Earth, respectively. The German physicist M. Planck (1858–1947) determined the distribution of radiated energy from a black body, an idealization for which

Climate Change in the Anthropocene.
DOI: https://doi.org/10.1016/B978-0-12-820308-8.00011-8

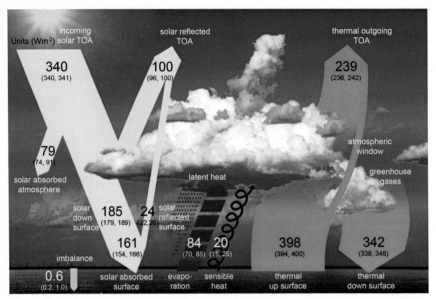

Figure 2.1 The global energy budget under present-day conditions. Numbers indicate energy fluxes magnitudes (Wm^{-2}) and numbers in parentheses include the range of values. *(Source with permission: Hartmann et al., 2013.)*

real objects are a reasonable approximation. The resulting black body curves of energy versus wavelength are similar to a skewed bell curve (Fig. 2.2). From Plank's work two generalizations were made. First, the energy flux E (Wm^{-2}) integrated over all wavelengths per unit time is proportional to T^4 (degrees Kelvin) and is given by: $E = \sigma T^4$ which is the Stefan–Boltzmann law and σ is the Stefan–Boltzmann constant (5.67×10^{-8} Wm$^{-2}\cdot$K^4). In terms of temperature, this can be written as: $T = (E/\sigma)^{1/4}$. The second generalization is that the maximum wavelength, λ_{max} emitted, is proportional to $1/T$ (Kelvin) which is Wien's law.

The sun at a temperature of about 5800 K (5527°C) emits radiation in the visible and ultraviolet spectral range with a maximum wavelength of 0.5 μm (μm $= 10^{-6}$ m). The Earth, being much cooler at about 288 K (15°C), emits radiation at longer wavelengths in the infrared spectral region with a maximum wavelength of 10 μm (range of 4–40 μm). In the visible light spectral range, much of the sun's energy passes through the atmosphere and reaches the Earth's surface as discussed above (Fig. 2.1). On the other hand, in the infrared spectral region a substantial amount of the energy emitted by the Earth is trapped by greenhouse gases such as water vapor, CO_2, CH_4, and N_2O (see below). Fig. 2.2 shows the spectral regions where

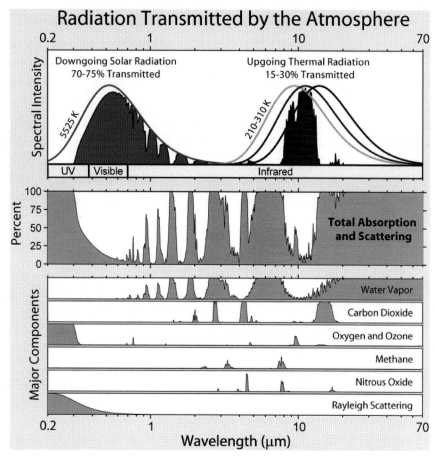

Figure 2.2 Radiation transmitted by the atmosphere according to wavelength (μm). Top panel – solar incoming radiation in the visible and ultraviolet range and outgoing radiation in the infrared region. Center panel – windows in the atmosphere for incoming and outgoing radiation (light areas). Lower panel – absorption of various gases (dark areas). Note water vapor and carbon dioxide on the right hand side. *(Source: NASA.)*

these gases prevent much of this radiation from reaching outer space resulting in the heating of the Earth's atmosphere and surface. This is the greenhouse effect. That these gases trapped infrared radiation was first shown by the Irish scientist John Tyndall in experiments undertaken in London in 1859. The Swedish scientist Svante Arrhenius (1859–1927) quantified these results in 1896 and calculated that a doubling of CO_2 would result in a 5°C to 6°C rise, in line with modern estimates. Table 2.1 summarizes the contrasts between the Sun and the Earth as radiating bodies.

Table 2.1 Contrast between the Sun and the Earth as black body radiators.

	T (K)	λ_{max} (µm)	Spectrum	E (Wm^{-2})
Sun	5800	0.5	Visible/UV	10^8
Earth	288	10	infrared	390

Source: author.

Ozone (O_3) blocks much of the incoming solar radiation in the ultraviolet region (Fig. 2.2) which protects life on Earth from harmful radiation. Ozone is also a greenhouse gas in the infrared region. There is, however, an ozone and water vapor "window" at about 10 µm allowing radiation to pass out of the atmosphere. Otherwise, water vapor, carbon dioxide, ozone, and nitrous oxide block radiation in the infrared region (4 to 40 µm) leading to global warming (Fig. 2.2).

If there were no greenhouse gases in the Earth's atmosphere the equilibrium temperature (T_e) of the Earth's surface and atmosphere can be calculated by equating the incoming solar radiation at the top of the atmosphere and the outgoing radiation from the Earth using the following expression (Tuckett, 2016):

$$T_e = \left[\frac{E(1-A)}{4\delta} \right]^{1/4} \tag{2.1}$$

where A is the average albedo of the Earth (0.32) and E equals 1360 Wm^{-2} and the value of σ is given above. This expression is just a restatement of the Stefan-Boltzmann law above applied to the Earth as a black body radiator. Using these values, T_e equals 253 K or negative 20°C. Without pre-industrial greenhouse gases the Earth would be a frozen snowball and life might never have begun due to the absence of liquid water. The Earth is warmer by 35°C than it would be otherwise without greenhouse gases. The effect of anthropogenic greenhouse gas concentrations is to increase the surface temperature so that the Earth emits greater amounts of long wave radiation (causing cooling), in an attempt to bring the radiative imbalance back into equilibrium.

2.3 Radiation forcing

The concept of radiative forcing (RF) is important because most other quantitative estimates of climate change are based on this quantity, including projected global mean surface temperature estimates. Radiative forcing addresses the radiative imbalance in the atmosphere due to both natural and

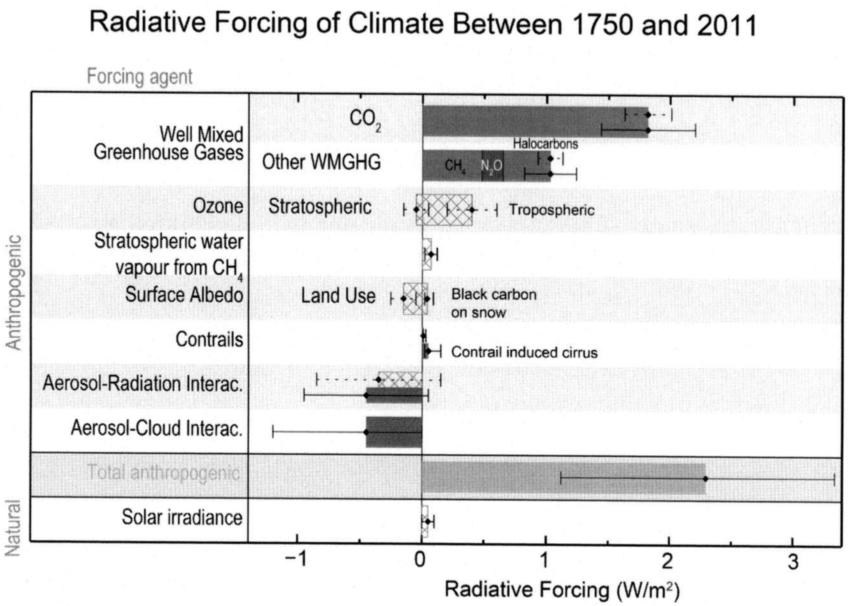

Figure 2.3 Bar chart of radiative forcing (W/m²) for the period 1750–2011 for different forcing agents. Cross hatch refers to effective radiation forcing. *(Source with permission: IPCC-AR5, Myhre et al., 2013.)*

anthropogenic activities from about 1750 AD to modern times (Fig. 2.3). It is defined as the net radiative flux (Wm^{-2}) averaged over a period of time at the tropopause (the boundary between the troposphere and stratosphere) after allowing for the stratosphere temperature to adjust to equilibrium while holding the surface and troposphere variables constant (Myhre et al., 2013). The RF, however, is not the *response* of the climate to the irradiance change – it is the irradiance change itself. The response of the climate to radiative forcing is a dis-equilibrium process on a range of time scales.

A modified RF, introduced in IPCC-AR5 (Myhre et al., 2013), is the effective radiative forcing: ERF = RF + rapid adjustments in RF. Factors that can cause rapid changes in RF include changes in cloud cover, vegetation, aerosols and snow cover. Not included in the ERFs are extent of sea ice or sea surface temperatures, which are held fixed by definition. The ERF is defined as the net downward radiative flux after allowing for troposphere temperatures, water vapor, and clouds to adjust, but with surface conditions held constant. Differences between RF and ERF are generally small, but ERF calculations have the potential to better model future climate changes particulary with regard to cloud-aerosol interactions (Fahey et al., 2017). The essential input to computer models to calculate RF for greenhouse gases is the spectroscopic properties of the individual gases.

Natural drivers of climate change include the El Niño-Southern Oscillation (ENSO), volcanic eruptions, and changes in solar irradiance, Earth orbital variations (i.e., Milankovitch cycles) and chemical weathering of rocks. Earth orbital cycles and chemical weathering are slow processes on millenia and longer timescales. Solar irradiance typically follows an 11 year cycle. The IPCC-AR5 report (Myhre et al., 2013) and the NCA4 report (Fahey et al., 2017) attribute a very small RF to solar irradiance (Fig. 2.3). The effect of large explosive volcanoes can last up to 2 years and have climate effects due to emissions of SO_2, leading to sulfuric acid deposition and cloud-aerosol interactions, resulting in a negative RF (i.e. cooling). This small effect is not shown in Fig. 2.3.

Anthropogenic climate drivers include the well mixed greenhouse gases, the principal ones being carbon dioxide (CO_2), methane (CH_4) and nitrous oxide (N_2O). They show globally well mixed concentrations with small interhemispheric gradients. They have lifetimes in the atmosphere from a decade (e.g., methane) to centuries and longer (carbon dioxide). They have all increased substantially in the Anthropocene (i.e. since 1750) and the current levels of CO_2 (420 ppm in 2021) are all well above concentrations seen in ice cores over the last 800,000 years (Jansen et al., 2007). From Fig. 2.3 it can be seen the RF contribution of these gases is large (including the halocarbons) and equals 2.8 Wm^{-2}.

2.4 Global warming potential

The concept of global warming potential (GWP) was introduced in IPCC-AR1 (Shine et al., 1990) to compare the greenhouse effects of different greenhouse gases relative to a reference gas, normally taken as carbon dioxide. Under this definition, CO_2 would have a GWP value of 1. It is defined as:

$$GWP = \frac{\int_0^a a_i \, c_i \, dt}{\int_0^a a_{co_2} \, c_{co_2} \, dt} \qquad (2.2)$$

where a_i is the instantaneous radiative forcing (RF) due to increase in concentration of gas i, c_i is the concentration of the gas remaining after time t and n is the number of years over which the calculation is performed. The corresponding values for the reference gas (CO_2) are in the denominator. It is assumed the gas is released instantaneously into the atmosphere and that its concentration decreases over time and during this time it produces greenhouse warming. If it is assumed gas i is removed only in proportion to its concentration then:

$$c_i(t) = e^{-t/\tau} \qquad (2.3)$$

where τ is the average lifetime of the gas in the atmosphere (Lashof and Ahuja, 1990). The GPW concept is useful in estimating the cumulative radiative forcing of all the greenhouse gases together at the same time (see Chapter 7).

The assumption of exponential decrease appears to be fulfilled by N_2O and CFCs (chloroflurocarbons). The case of CO_2 is more complex as it does not have a well defined residence time since it is transferred between different reservoirs on different timescales (ocean, land, biota, atmosphere). Because of this, using carbon dioxide as the reference gas above may not be an optimal choice. Additional problems are the indirect effects on RF due to CH_4 -CO-OH reaction coupling. Methane emissions tend to increase tropospheric ozone and stratospheric water vapor possibly enhancing the indirect greenhouse effect of methane.

In physical terms, the GWP can be interpreted as an index of the total energy added to the climate system by a greenhouse gas relative to that of CO_2 (Myhre et al., 2013). These authors give the GWP value of methane as 84–86 over 20 years and 28–34 over 100 years. Nitrous oxide values over the same time periods are 264–268 and 265–298, respectively. The ranges in estimates are largely due to models that account for, or do not account for, climate-feedback mechanisms. These values are not that different from the early values given in the first IPCC report (Shine et al., 1990), but they are still relatively tentative. Some of the properties of the main greenhouse gases are described below.

2.5 Greenhouse gases

2.5.1 Carbon dioxide (CO_2)

Carbon dioxide in the atmosphere has increased by 50% since industrial times (~1750) from a pre-industrial level of 278 ppm to 420 ppm in 2021. Global fossil fuel emissions were 36.4 $GtCO_2$ in 2018. This value saw a decrease of 6% in 2020 due to the global pandemic. Carbon dioxide has several sources and sinks. Sources include burning fossil fuels (oil, gas, coal) and industry (cement manufacture) and land use change (e.g. biomass burning). Sinks include the oceans, the atmosphere, and land vegetation (Fig. 2.4). A number of processes remove CO_2 from the atmosphere including uptake by the oceans, uptake from the land and rock weathering. The

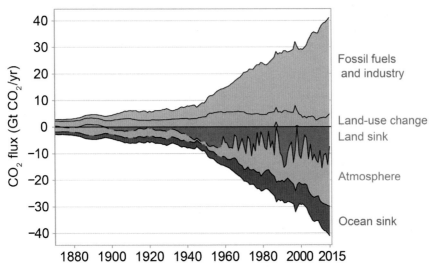

Figure 2.4 Sources (above zero) and sinks (below zero) of CO_2 (Gt/year) for the period 1870–2015. *(Source: NCA4, Fahey et al., 2017.)*

latter process acts as a sink by forming carbonic acid (H_2CO_3) from CO_2 and H_2O. These processes operate on different timescales spanning decades to millennia so that the lifetime of CO_2 in the atmosphere is longer than that of the other GHG and is difficult to specify uniquely. Seasonal variations also occur due to photosynthesis as shown by the Keeling curve at Mauna Loa, Hawaii (Keeling, 1960; http://ersl/gmd/ccgg/trends). In addition to buried fossil fuels, there are other large carbon reservoirs: the oceans, vegetation, soils, and permafrost. Carbon dioxide reached 420 ppm at Mauna Loa in 2021 and the average rate of increase over the period 2016–2018 was 2.6 ppm/yr. At this rate the concentration would reach 490 ppm by 2050. Fig. 2.5 shows the CO_2 concentrations (ppm) over the past one thousand years based on cores from Law Dome in Antarctica (Etheridge et al., 1998) with more recent data from Dlugokencky et al. (1994).

2.5.2 Nitrous oxide (N_2O)

Growth rates and RF values for N_2O are lower than both CH_4 and CO_2 (Fig. 2.2). Nitrous oxides' growth rate has been about 20% since preindustrial times from 260 ppb to 320 ppb (parts per billion) in 2018 (Fig. 2.5). Since the 1950s its growth rate is largely due to the use of agricultural fertilizers and motor vehicle exhaust and some manufacturing processes. Other sources are soils and the oceans. The main sink is by photochemical destruction

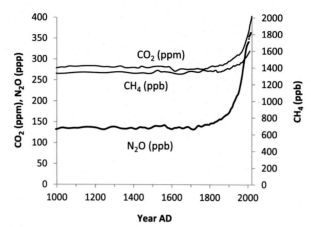

Figure 2.5 Concentrations of CO_2, CH_4, and N_2O over the past thousand years in ice cores from Law Dome, Antarctica. *(Source: NOAA.)*

in the stratosphere that produces nitrogen oxides (NO_x) and reduces the concentration of ozone (O_3). Nitrous oxide is part of a larger global budget of reactive nitrogen involving ammonia (NH_3), NO_x, and other compounds. Annual emissions are estimated at 18 MtN (million metric tons nitrogen), 40% of which is anthropogenic. Its lifetime in the atmosphere is estimated at 120 years and it is a very strong greenhouse gas with a GWP value in the range 265–300 (Fahey et al., 2017).

2.5.3 Methane (CH_4)

The globally averaged methane concentration in the atmosphere was about 722 ppb in 1750 and 1866 ppb in 2019, a 250% increase over pre-industrial values (Fig. 2.5). Total global emissions were 556 ± 56 $MtCH_4$ in 2011 with 60% from anthropogenic sources. It is a much stronger greenhouse gas than carbon dioxide (with a GWP $\approx$ 85; see above) but because of its lower concentration it has a lower RF value (Fig. 2.2). The lifetime of methane in the atmosphere is about a decade and it has several natural and anthropogenic sources and sinks. It's most important sink is by reaction with the hydroxyl radical (OH) in the troposphere (< 10 km). In the stratosphere (>10 km) it reacts with OH to produce water vapor, which itself is a strong greenhouse gas, causing an indirect or secondary positive ERF (Fig. 2.2).

Methane is produced in anaerobic (oxygen deficient) environments and natural sources include wetlands such as bogs, swamps and thawing permafrost. An additional potential very large source is frozen clathrates (methane molecules encased in ice molecules) in frozen continental margin sediments (Kopp et al., 2017). During the interglacial (warm) periods of the last ice age, spikes in concentration of methane in ice cores (ranging from 350 ppb to 700 ppb) are attributed to emission from tropical wetlands since northern wetlands sources were likely still frozen at that time (Chapppellaz et al., 1993). Anthropogenic sources include rice paddies, farm animals, coal mining, oil and gas extraction, landfills, and biomass burning. These sources would tend to increase as the human population increased and as farming and industrialization became more intense at the beginning of the Anthropocene.

2.5.4 Halocarbons

Halocarbons are another group of well mixed greenhouse gases that are synthetic (i.e. manufactured). They include chlorofluorocarbons (CFCs), hydrochlorofluorocarbons (HCFCs), hydrofluorocarbons (HFCs), perfluorocarbons (PFCs), and sulphur hexafluoride (SF_6) (Fahey et al., 2017). In the industrial era most of these gases have anthropogenic sources. Chlorofluorocarbons (CFCs) saw expanded use as refrigerants in the mid-twentieth century. Under the Montreal Protocol , the production of CFCs was regulated because of their destruction of ozone in the stratosphere and by the 1990s CFCs saw a decline. The halocarbon greenhouse gases collectively show a wide range in GWPs, and a wide range in atmospheric concentrations spatially and temporally. Together they have a small greenhouse effect compared to the other greenhouse gases (Fig. 2.2). In an amendment to the Montreal Protocol in 2016, other countries and the United States, agreed to phase down production of HFCs (Fahey et al., 2017).

2.5.5 Ozone (O_3)

Ozone has a high temporal and spatial variability making RF calculations difficult. Two types of ozone need to be considered: tropospheric ozone (<10 km) yielding a positive RF of approximately 0.4 Wm^{-2} (range 0.2–0.6 Wm^{-2}) and stratospheric ozone. Tropospheric ozone is produced by photochemical reactions with precursor species, both anthropogenic and natural species such as CH_4, NO_x, CO, and other organic compounds. Stratospheric ozone is destroyed by the halogens (e.g., chlorine and bromine), which in turn are produced by natural and anthropogenic precursors. Stratospheric ozone increased substantially in the 1950s until the 1990s when it declined. Depletion of stratospheric ozone is most notable in the polar regions and its depletion has a small negative RF (Fig. 2.2). The control of CFC production under the Montreal Protocol has moderated ozone depletion in the stratosphere (the so-called ozone hole over Antarctica).

2.5.6 Water vapor

The main control on water vapor in the atmosphere is temperature, with humidity increasing by 7% for every degree (°C) in warming, as predicted by the Clausius-Claperyon relation and confirmed by satellite observations over the oceans (Myhre et al., 2013; Fahey et al., 2017). Anthropogenic input of water vapor, for example by evaporation from irrigation or power plant cooling towers, is negligible. Water vapor has the largest greenhouse effect on the atmosphere, but it is not treated as an anthropogenic greenhouse gas since as mentioned temperature is the main determinant of its concentration. In addition, it has three phases (solid, liquid, and gas) unlike other anthropogenic GHGs. As the atmosphere warms its water vapor content increases, resulting in a positive feedback. Also, as noted already, breakdown of anthropogenically produced methane in the stratosphere (>10 km) produces water vapor which also has a positive radiative forcing. So, while water vapor is the most powerful greenhouse gas it is accounted for in computer models as a feedback gas and also by its indirect production in the atmosphere. Stratospheric water vapor is produced by transport from the tropical troposphere and by oxidation of methane and by injection from volcanic plumes.

2.6 Aerosols

Aerosols are the most complex of all the anthropogenic radiative forcers (Boucher et al., 2013; Fahey et al., 2017). They can be classified by composition: sulfate, black carbon, organic, nitrate, dust and sea salt. Black carbon (soot) may be due to either fossil fuel or biomass burning. They have lifetimes of days to weeks and they have high spatial and temporal variability, and they are precipitated by rain. Three types of aerosol interactions due to anthropogenic activities can be identified. The first is aerosol-radiation interaction where both long wave and short wave radiation is scattered (termed a direct effect) producing a negative (cooling) RF effect. The second is aerosol-cloud interaction called the cloud-albedo or indirect effect. This also produces a negative RF effect due to increased cloud albedo. The third type is light-absorbing aerosols, namely black carbon, which produces a positive RF especially on snow and ice where the albedo is decreased producing a warming of the surface. This in turn increases cloud cover, reducing the positive RF effect of the light-absorbing aerosol by 15%. It is only relatively recently these complex effects have been quantified (Fig 2.2).

2.7 Climate response

Climate change occurs in response to ERFs (see effective radiative forcings above) and involves both positive and negative feedbacks on a range of time scales (e.g., days to centuries). Of interest are two types of response, one short-term the other long-term. The short-term response is called the transient climate response (TCR) which is defined as the change in surface temperature due to a doubling of CO_2 at a rate of 1% per year. This is the climate response at a given time due to a continously evolving forcing. The longer term response is called the equilibrium climate sensitivity (ECS) and it is the total climate response for a given forcing. It is defined as the equilibrium surface temperature due to doubling of CO_2. In this case climate models may have to simulate thousands of years of climate before equilibrium is reached. Based on different models, values for ECS are in the range 1.5°C–4.5°C and are greater than values for TRC (1°C–2.5°C), indicating that long-term feedbacks are significant and positive. The wide range in ECS values are thought to be due to the variable effects of clouds (Knutti and Rugenstein, 2015). In addition to climate models, other sources of information include measurements of ocean temperatures in the industrial era, paleoclimate archives, and forcings due to recent volcanic eruptions.

While TCR and ECS specifically relate to a doubling of CO_2, a more general parameter is the climate sensitivity parameter, λ (°C/Wm^{-2}), defined as the change in the equilibrium annual mean global surface temperature (ΔT) due to a unit change in RF:

$$\Delta T = \lambda RF \qquad (2.4)$$

Different forcings from different sources are assumed to be linearly additive, although this assumption has been questioned (Knutti and Rugenstein, 2015). The value of the sensitivity parameter cannot be measured but is an emerging property of the system. Estimates can be made from past climates assuming equilibrium warming and estimated radiative forcings.

Eq. 2.4 is highly sensitive to feedbacks. Feedbacks (see Box 2.1) are themselves assigned RFs since they affect either the greenhouse effect or the Earth's albedo. Based on models for the industrial era a value for λ is approximately 0.5°C/Wm^{-2}. In the absence of feedbacks this value is 0.3°C/Wm^{-2}

indicating positive feedbacks amplify the industrial era forcings (Fahey et al., 2017). Feedbacks are the largest source of uncertainty for λ, which accounts for near-term feedbacks such as response to clouds, the carbon cycle, aerosols, ocean circulation and precipitation changes. Geographic and seasonal variations of forcings increase uncertainty as do spatial and temporal variations in anthropogenic concentrations and emissions. Examples of inhomogeneous short-term forcings include aerosols, tropospheric ozone, contrails, and land cover changes. In the case of projecting regional climate change, as opposed to global, there is low confidence in the response to forcings because of difficulty in distinguishing between forcing responses and inherent climate system variability.

BOX 2.1 Quantifying feedbacks

Eq. 2.4 in the main text defined the sensitivity parameter λ so that the temperature increase (ΔT_o) due to a radiative forcing (RF) is:

$$\Delta T_o = \lambda (RF) \tag{2.1.1}$$

Here ΔT_o is taken to be a reference system with no feedbacks. If a feedback, either positive or negative, is included in the system the simplest representation is that the feedback forcing is linearly proportional to the system response namely $\alpha_1 \Delta T$, where α_1 is a constant (Roe, 2009). The feedback forcing is added to the original forcing:

$$\Delta T = \lambda (RF + \alpha_1 \Delta T) \tag{2.2.2}$$

Similarly, assuming the effects of several feedbacks are linearly additive we can write:

$$\Delta T = \lambda (RF + \alpha_1 \Delta T + \alpha_2 \Delta T + \alpha_3 \Delta T) \tag{2.2.3}$$

In (Eq. 2.2.3) the value of λ remains the same as in (Eq. 2.2.2); it is the radiative forcing term that changes. Solving for ΔT in (Eq. 2.2.2):

$$\Delta T = \left[\frac{\lambda RF}{1 - \alpha_1 \lambda} \right] \tag{2.2.4}$$

The system gain, G, is the factor by which the system temperature has increased or decreased compared to the reference system (Eq. 2.2.1):

$$G = \left[\frac{\Delta T}{\Delta T_o} \right] \tag{2.2.5}$$

The reference system might be Earth with pre-industrial GHG concentrations and the system with feedbacks might represent a post-industrial doubling of CO_2. The feedback factor, f, is defined as the fraction of the output fed back into the input

(continued)

Box 2.1 Quantifying feedbacks— *cont'd*
(see Fig. 2.1):

$$f = 1 - \alpha 1 \lambda \qquad (2.2.6)$$

Combining (Eqs. 2.2.4–2.2.6) it is easy to show that:

$$G = \left[\frac{1}{1-f} \right] \qquad (2.2.7)$$

This relationship, between the feedback factor and the system gain, provides some interesting insights. For $f = 0$ (no feedbacks), the gain, G ($\Delta T/\Delta T_o$), is unity as expected. For $f > 0$ we have positive feedback indicating additional heating. For example when $f = 0.5$, $G = 2$ so that the system output is doubled. For $f < 0$ we have negative feedback; when $f = -0.5$, the system output is 0.66 indicating cooling. Due to the exponential shape of the $1/(1 - f)$ function, negative and positive feedbacks are not symmetrical.

The assumption of a constant climate sensitivity parameter (λ) is only an approximation and more recent climate models show that it is a function of temperature, time and forcing strength (Knutti and Rugenstein, 2015). These authors show that model outputs during the transient climate response indicate that λ decreases strongly within the first hundred years but as equilibrium is approached it levels off so that a constant λ assumption is reasonable in some circumstances for longer time scales (e.g. several centuries).

2.8 Feedbacks

2.8.1 Short-term feedbacks

Short-term feedbacks include albedo changes due to seasonal snow and ice extent, water vapor, and changes in the atmosphere vertical temperature gradient (called the lapse rate) and clouds.

As noted above, when the Earth's surface increases in temperature due to an increase in incoming radiation, the Earth acts as a blackbody radiator and will emit more long wave (infrared) radiation through a window in the atmosphere (at about 10 μm wavelength) to outer space. This is a negative feedback, resulting in cooling. It serves to restore radiative balance at the top of the atmosphere bringing the energy budget toward equilibrium. This feedback only partly offsets the original heating but is nevertheless the most important negative feedback in the climate system, sometimes called the Planck effect. According to IPCC-AR5 (Myhre et al., 2013) it has a RF value of negative 3.4 Wm^{-2} and NCA4 (Fahey et al., 2017) provides a somewhat similar value of negative 3.2 Wm^{-2}.

Water vapor provides a powerful positive feedback as the vapor content of the atmosphere increases with increasing temperature. An initial temperature increase therefore produces more moisture in the atmosphere which itself enhances the greenhouse effect. Atmosphere-ocean global climate models (AOGCMs; see Chapter 3) indicate an RF value of 0.16 $\pm$ 0.3 W/m^{-2} per 1°C warming. Water vapor more than doubles the direct RF effect of increasing CO_2.

The IPCC-AR5 assessment report (Myrhe et al., 2013) suggests that water vapor and the lapse rate feedbacks (the vertical temperature gradient in the atmosphere) should be considered together. Heating due to water vapor causes more heating at the top of the troposphere (where it is colder), thereby reducing the lapse rate. This higher altitude heating causes more outgoing radiation – a negative feedback, estimated at negative 0.6 $\pm$ 0.4 Wm^{-2}. Overall, however, the water vapor lapse rate feedback is significantly positive (Fahey et al., 2017).

Cloud-radiation and cloud-aerosol interactions are complex and represent one of the largest uncertainties in climate model projections (Boucher et al., 2013). Feedbacks vary from positive to negative and depend on cloud type, altitude, latitude, and aerosol type. The IPCC-AR5 report (2013) devoted a lengthy chapter to this topic for the first time. The NCA4 report (Fahey et al., 2017) provides a shorter overview which is summarized below.

Cloud-radiation interactions fall into both a short-wave and a long-wave response. Increase in cloudiness increases scattering of sunlight, thereby increasing the Earth's albedo and cooling the surface (short-wave effect). The global ERF for this effect is estimated at negative 50 Wm^{-2} the largest RF in the entire climate system. This effect is larger over dark surfaces (e.g., oceans) compared to higher albedo landscapes such as ice or deserts. For long-wave radiation, cloudiness traps infrared which warms the surface with an estimated positive ERF of 30 Wm^{-2} leading to a net cooling of negative 20 Wm^{-2}. The magnitude of both effects varies with cloud type and location.

For low thick altitude clouds (e.g., stratus, stratocumulus) it appears the short-wave effect dominates producing a net cooling due to high albedo of the upper cloud surfaces. This effect dominates at mid-latitudes. In the case of thin high altitude clouds (e.g., cirrus) the long-wave effect dominates producing warming. Satellite observations over the period 2001–2011 indicate a global net cooling due to clouds (Boucher et al., 2013).

Table 2.2 Typical surface albedo values.

Surface	Albedo %
Snow/ice	40–90
Water (high sun)	3–10
Desert	20–45
Grass	16–26
Forest	5–20

Source: O'Hara, 2014.

2.9 Albedo feedbacks

Snow and ice are highly reflective to solar radiation relative to other surfaces (Table 2.2).

Atmospheric warming causes melting of snow, sea ice, and land ice. In the case of sea ice this melting leads to heating of the seawater on account of its lower albedo which in turn causes further ice melting – an important positive feedback in the Arctic. It is estimated the albedo-ice feedback contributes 0.27 ± 0.06 Wm^{-2} per 1°C warming to the RF (Fahey et al., 2017). Cooling during an ice age, for example, will cause the reverse effect, also a positive feedback. Deposition of black carbon (soot, from fossil fuel or biomass burning) on ice or snow decreases the albedo and absorbs incoming radiation causing more melting than would otherwise occur.

Continental ice sheets such as Greenland and Antarctica have ice shelves on their margins that help stabilize the edges of the ice sheet by retarding discharge into the oceans. If ice shelves melt, possibly due to ocean heating, then the continental ice sheet can undergo increased mass loss. Significant input of melt water in the oceans can change its salinity and temperature and therefore disrupt the ocean "conveyor" belt circulation. It has been hypothesized that calving of ice off the Laurentian ice sheet during the last Ice Age led to the shut-down of the warm Gulf Stream in the North Atlantic producing a short period of extreme cold in the region (called the Younger Dryas, 12,000–11,500 years ago) and this effect is visible in Greenland ice cores. The calving of the icebergs into the North Atlantic also left a record of coarse sediment in deep sea cores in the region at the same time. These effects indicate complex feedbacks and interactions in the cryosphere–ocean system.

2.9.1 Long-term feedbacks

The carbon cycle is a series of carbon reservoirs connected by exchange fluxes between reservoirs (Ciais et al., 2013; Fahey et al., 2017). The reservoir turnover time or residence time is the reservoir mass divided by the exchange flux (GtC/yr.) and range from a few years to millennia for major reservoirs including land vegetation, soils and ocean reservoirs (shallow and deep ocean). These are referred to as the fast exchange domain). A second slow domain operates on a turnover time of 10,000 years or longer and includes stores of carbon in rocks and sediments that exchange with the fast domain by chemical weathering, volcanic activity, erosion, and sedimentation on the ocean floor.

The cycling of carbon through the climate system is an important long-term feedback that affects the concentration of CO_2 in the atmosphere. The atmosphere in 2020 had a carbon dioxide concentration of 417 ppm corresponding to a store of carbon of 884 GtC (1 ppm is equivalent to 2.12 GtC). The global mean atmosphere CO_2 concentration is determined by the emissions due to burning fossil fuels and cement manufacture, wild-fires, permafrost thawing, balanced by sinks such as the terrestrial biosphere and the oceans. Over the past decade about a third of anthropogenic CO_2 has been taken up by the terrestrial environment and a quarter by the oceans through photosynthesis and direct uptake by surface water. The capacity of the terrestrial environment and the oceans to continue CO_2 uptake is highly uncertain and is dependent on climate change in the future (Chapter 3).

2.10 Ocean chemistry, ecosystems, and circulation

The oceans play a significant role in controlling the amount of GHG (CO_2, water vapor, N_2O) and heat in the atmosphere. Most of the net energy increase in the climate system due to anthropogenic RF is heat in the oceans with about 60% confined to the upper 700 m. About 93% of excess heat in the climate system over the past 50 years is found in the oceans (Rhein et al., 2013). This is because of the large heat capacity and mass of seawater compared to the atmosphere. Moreover, the shallow oceans connect to the deep ocean through circulation. The ocean also has a low albedo making it a better absorber of solar radiation. Ocean warming and changes in ocean stratification (involving temperature or salinity) alter ocean circulation and biological productivity and therefore CO_2 uptake. Marine ecosystems take CO_2 from the atmosphere through photosynthesis. Marine plants such as phytoplankton have a net primary production (NPP) of about 50 ± 28 GtC/yr. A substantial fraction of this is sequestered in the deep ocean where it is isolated from the atmosphere on a long-term basis (referred to as a biological pump). The oceans contain about 50 times more inorganic carbon than the atmosphere.

Ocean uptake of carbon is computed from the observed difference in the partial pressure of CO_2 (pCO_2) across the air-water interface:

$$\Delta pCO_2 = pCO_{2\,water} - pCO_{2\,air} \qquad (2.5)$$

The paucity of ΔpCO_2 measurements both spatially and temporally and the associated large uncertainty precludes detecting any global trends in the ocean uptake flux. Trends at the Mauna Loa station, Hawaii, however, do show increases in $pCO_{2\,water}$ over the period 1990–2012, consistent with the monotonic atmospheric increase in CO_2 concentration at that site (Rhein et al., 2013). Irregularities in the oceanic partial pressure may be caused by changes in biological productivity or changes in physical conditions (e.g., due to El Niño and La Niña).

There are several feedbacks involved in the ocean-atmosphere-cryosphere system. Climate-caused changes in ocean temperature, circulation, and stratification alter phytoplankton NPP and therefore the carbon sink. In addition, increase in the ocean acidity may change the NPP which may also alter CO_2 uptake. The ocean also alters the hydrological cycle, becoming more intense with warming due to the higher moisture content of the atmosphere. Most ($\sim$80%) of evaporation and precipitation takes place over the oceans. Evaporation is controlled by wind stress and water temperature. The historical record of precipitation and evaporation over the oceans is poor. However, salinity (grams of dissolved solids in one kg of saltwater) is a good proxy acting as a rain gauge. High salinity regions reflect areas where evaporation exceeds precipitation and low salinity regions where precipitation and runoff exceeds evaporation. Freezing and melting of ice and ocean circulation also influence salinity. The global array of Argo floats ($\sim$3000 in all) deployed in the 2000s has greatly extended salinity (and temperature) measurements spatially and temporally. Over the past several decades average salinity has decreased in the already low salinity Pacific and increased in the already high salinity Atlantic in the upper 500 m. In other words, the climate change trend intensifies the existing pattern of salinity. Salinity increases and decreases cause increase in stratification which alters circulation which in turn alters heat and CO_2 uptake. Increase in stratification inhibits surface mixing and deep water formation leading to weaker ocean circulation.

Increase in ocean temperature also accelerates melting of continental ice sheets, particularly in Antarctica. Freshwater input alters stratification which

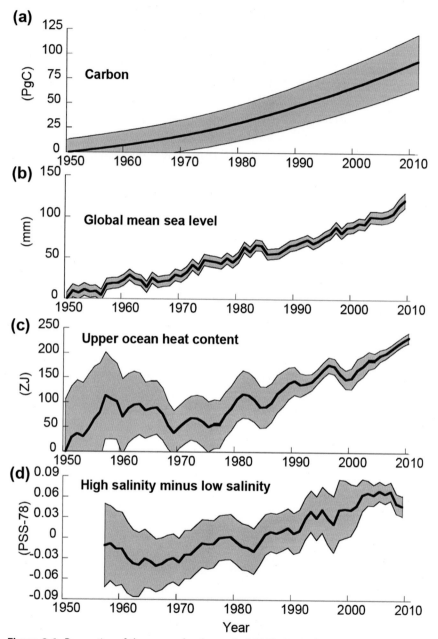

Figure 2.6 Properties of the ocean for the period 1950–2010. *(Source with permission: Rhein et al., 2013.)*

alters circulation again possibly affecting absorption of GHGs and heat, although this effect has not yet been quantified. Computer models project increases in stratification, decreasing NPP and decreasing sink for CO_2 via biological activity. This will increase atmospheric CO_2 and is therefore a positive feedback (Fahey et al., 2017). Fig. 2.6 summarizes several properties of the ocean for the period 1950–2010. The anthropogenic carbon dioxide content, mean sea level, heat content, and salinity differences all show marked increases. Ocean acidity (not shown) has also increased over the same period (Rhein et al., 2013). Clearly the oceans are undergoing major changes.

2.11 Permafrost

Wherever the ground remains frozen for two or more consecutive years is called permafrost. If cold ground temperatures continue for many years the permafrost thickens until equilibrium is reached where the outward interior heat flow balances the penetration of cold from above. Thicknesses in Alaska and Siberia of 600 m and 1500 m respectively have been reported, probably dating back to the last glaciation or earlier. The layer that thaws in the summer is referred to as the active layer. Such melting results in soil flows, landslides and subsidence causing problems for infrastructure such as roads and buildings. In most regions permafrost temperatures have increased over the past few decades by 1°C to 2°C and the number of days the oil exploration and forestry industries can operate in the year has decreased by 50% in Alaska in the past three decades (ACIA, 2004).

Permafrost contains large reservoirs of carbon in the form of organic materials. With warming this material is thawing releasing methane and carbon dioxide by microbial decomposition. This permafrost–carbon feedback is positive.

References

ACIA, 2004. Impacts of a wWarming Arctic: Arctic Climate Impact Assessment. Cambridge University Press, U.K. and New York http://www.acia.uaf.edu.

Boucher, O., Randall, D., Artaxo, P., Bretherton, C., Feingold, G., Forster, P., Kerminen, V.-M., Kondo, Y., Liao, H., Lohmann, U., Rasch, P., Satheesh, S.K., Sherwood, S., Stevens, B., Zhang, X.Y., 2013. Clouds and Aerosols. In: Stocker, T.F., Qin, D., Plattner, G.K., Tignor, M., Allen, S.K., Boschung, J., Nauels, A., Xia, Y., Bex, V., Midgley, P.M. (Eds.), Climate Change 2013: The Physical Basis. Contribution of Working Group 1 to the Fifth Assessment of the Intergovernmental Panel on Climate Change. Cambridge University Press, UK and New York.

Chappellaz, J., Blunier, T., Raynaud, D., Barnola, J.M., Schwander, J., Stauffer, B., 1993. Synchronous changes in atmospheric CH_4 and greenland climate between 40 and 8 kyr BP. Nature 266, 443–445.

Ciais, P., Sabine, C., Bala, G., Bopp, L., Brovkin, V., Canadell, A., Chhabra, R., DeFries, R., Galloway, J., Heimann, C., Jones, C., Le Quéré, R.B., Piao, S., Thornton, P., 2013. Carbon and Other Biogeochemical Cycles. In: Stocker, T.F., Qin, D., Plattner, G.K., Tignor, M., Allen, S.K., Boschung, J., Nauels, A., Xia, Y., Bex, V., Midgley, P.M. (Eds.), Climate Change 2013: The Physical Basis. Contribution of Working Group 1 to the Fifth Assessment of the Intergovernmental Panel on Climate Change. Cambridge University Press, UK and New York.

Dlugokencky, E.J., Steele, L.P., Lang, P.M., Masarie, K.A., 1994. The growth rate and distribution of atmospheric methane. J. Geophys. Res. 99, 17. doi:10.1029/94JD01245, 021–17,043.

Etheridge, D.M., Steele, L.P., Langenfelds, R.L., Francey, R.J., Barnola, J.-M., Morgan, V.I, 1998. Historical records from the Law Dome DE08, DE08-2 and DSS ice cores. Trends: A Compendium of Data on Global Change. Carbon Dioxide Information Analysis Center, Oak Ridge Laboratory, US Department of Energy, Oak Ridge, Tennessee, USA.

Fahey, D.W., Doherty, S.J., Hibbard, K.A., Romanou, A., Taylor, P.C., 2017. Physical drivers of climate change. In: Wuebbles, D.J., Fahey, D.W., Hibbard, K.A., Dokken, D.J., Stewart, B.C., Maycock, T.K. (Eds.), Climate Science Special Report: Fourth National Climate Assessment, 1. U.S. Global Change Research Program, Washington D. C., USA, pp. 73–113.

Hartmann, D.L., Klien Tank, A.M.G., Rusticucci, M., Alexander, L.V., Brönnimann, Charabi, Y., Dentener, F.J., Dlugokencky, E.J., Easterling, D.R., Soden, B.J., Thorne, B.W., Wild, M., Zhai, P.M., 2013. Observations: atmosphere and surface. In: Stocker, T.F., Qin, D., Plattner, G.K., Tignor, M., Allen, S.K., Boschung, J., Nauels, A., Xia, Y., Bex, V., Midgley, P.M. (Eds.), Climate Change 2013: The Physical Basis. Contribution of Working Group 1 to the Fifth Assessment of the Intergovernmental Panel on Climate Change. Cambridge University Press, U. K. and New York.

Jensen, E., Overpeck, J., Briffa, K.R., Duplessy, J.-C., Joos, F., Masson-Delmotte, V., Olago, D., Ott-Bliesner, B., Peltier, W.R., et al., 2007. Paleoclimate. The physical science basis. In: Solomon, S., Qin, D., Manning, M., Marquis, M., Averyt, K., Tignor, M.M.B., Miller, H.L., Chen, Z. (Eds.), Contributions of Working Group I to the fourth Assessment Report of the Intergovernmental Panel on Climate Change. Cambridge University Press, UK and New York.

Keeling, C.D., 1960. The concentration and isotopic abundances of carbon dioxide in the atmosphere. Tellus 12, 200–203.

Knutti, R., Rugenstein, M.A.A., 2015. Feedbacks, climate sensitivity and the limits of linear models. Phil. Trans. Roy. Soc. A 373, 20150146.

Kopp, R.E., Hayhoe, K., Easterling, D.R., Hall, T., Horton, R., Kunkel, K.E., LeGrande, A.N., 2017. Potential surprises - compound extremes and tipping elements. In: Wuebbles, D.J., Fahey, D.W., Hibbard, K.A., Dokken, D.J., Stewart, B.C., Maycock, T.K. (Eds.). Climate Science Special Report; Fourth National Climate Assessment, 1. U.S. Global Change Research Program, Washington D. C., USA.

Lashof, D.A., Ahuja, D.R., 1990. Relative contributions of greenhouse gas emissions to global warming. Nature 344, 529–531.

Myhre, G., Shindell, D., Bréon, F.-M., Collins, W., Fuglestvedt, J., Huang, J., Koch, D., Lamarque, J.-F., Lee, D., Mendoza, B., Nakajima, T., Robock, A., Stephens, G., Takemura, T., Zhang, H., 2013. Anthropogenic and natural radiative forcing. In: Stocker, T.F., Qin, D., Plattner, G.K., Tignor, M., Allen, S.K., Boschung, J., Nauels, A., Xia, Y., Bex, V., Midgley, P.M. (Eds.), Climate Change 2013: The Physical Basis. Contribution of Working

Group 1 to the Fifth Assessment of the Intergovernmental Panel on Climate Change. Cambridge University Press, UK and New York.

O'Hara, K., 2014. Earth Resources and Environmental Impacts. Wiley, Hoboken, New Jersey.

Rhein, M., Rintoul, S.R., Aoki, S., Campos, E., Chambers, D., Feely, R.A., Gulev, S., Johnson, S.A., Kostianoy, A., Mauritzen, C., Roemmich, D., Talley, L.D, Wang, F., 2013. Observations; Ocean. Climate change. In: Stocker, T.F., Qin, D., Plattner, G.K., Tignor, M., Allen, S.K., Boschung, J., Nauels, A., Xia, Y., Bex, V., Midgley, P.M. (Eds.), The Physical Basis. Contribution of Working Group 1 to the Fifth Assessment of the Intergovernmental Panel on Climate Change. Cambridge University Press, UK and New York.

Roe, G., 2009. Feedbacks, timescales and seeing red. Ann. Rev. Earth Planet. Sci. 37, 93–115.

Shine, K.P., Derwent, R.G., Wuebbles, D.J., Morcrette, J.-J., 16 others, 1990. Radiative forcing of climate. In: Houghton, J.T., Jenkins, G.J., Ephraums, J.J. (Eds.), Climate Change: The Intergovernmental Panel on Climate Change Working Group 1. Cambridge University Press, UK and New York, pp. 45–68.

Tuckett, R.P., 2016. The role of atmospheric gases. In: Letcher, T.M. (Ed.), Climate Change, Observed Impacts on Planet Earth, 2nd ed. Elsevier, Amsterdam.

CHAPTER 3

Evaluation of climate model performance

3.1 Introduction

Computer climate models are the primary tool for making projections of climate change and they range from simple idealized models to models of intermediate complexity (EMICs) to comprehensive global circulation models (GMCs). More sophisticated models such as earth system models (ESMs) also include biogeochemical cycles, such as the carbon cycle, and they are used in the fifth IPCC-AR5 assessment report (Flato et al., 2013). Atmosphere-ocean general circulation models (AOGCMs) however continue to be used with new improvements. The coupled model Intercomparison project phase 3 (CMIP3) and phase 5 (CMIP5), are two internationally coordinated simulation projects that explore different model parameters and the range of results they produce. These models simulate many aspects of the climate including: temperature of the atmosphere, land, and oceans, precipitation, winds, clouds, sea-ice extent, and ocean currents. The better these models successfully simulate historical climate observations and paleoclimates, the more confidence we can have in their ability to predict future climate scenarios.

Climate models play an important role in studying climate change and they can be used to address a variety of issues: for example, simulation of mean climate on a variety of spatial scales (global to regional) and time scales from decades, centuries, to millennia either in the historical record or the paleoclimate record and also for future projections. They can also be used to provide insight into key climate processes, such as feedbacks. These models are commonly used to estimate emission scenarios required for climate stabilization targets in the future (IPPC-AR5, 2013). For example, if climate is to be stabilized at 2.0°C or less above pre-industrial levels by the end of the century (e.g., the Paris Accords of 2015) what GHG emission scenarios should be followed?

Climate Change in the Anthropocene.
DOI: https://doi.org/10.1016/B978-0-12-820308-8.00007-6

All climate models are based on scientific laws such as conservation of energy, mass, and momentum. The building and implementation of a climate model commonly involves the following steps:

- Express the systems' physical laws in mathematical terms, which require theoretical and observational input.
- Implement these mathematical expressions on a computer. Numerical methods are commonly used to solve complex equations without analytical solutions. The programs are normally implemented on a three-dimensional grid (e.g., longitude, latitude, and elevation) with variable resolution.
- Parameterization (or tuning) involves implementing conceptual models (e.g., diffusion versus convection in the ocean or a combination of both) and the input of numerical parameters that reproduce observational data. The processes that are tuned are commonly on a sub-grid scale, mainly due to a lack of data on that scale. Quantities that are "tuned" cannot be used to evaluate the success of a model, which would be a type of circular reasoning.

State-of-the-art climate models require significant supercomputer resources and the resolution (spatial and temporal) of the model must be balanced against the increased cost. The term ensemble is commonly used to describe a parallel group of model experiments, and differences in results give an estimate of uncertainty. A large number of computer simulations, using a single model with different initial conditions, allow climate variability to be explored. Alternatively, different model simulations can be done allowing model differences to be explored. In short, numerous simulations can be undertaken with a single model with different parameters or alternatively different models can be explored to explore these same parameters. Two fundamentally different types of model experiment can be undertaken. In an equilibrium climate model, the experiment is allowed to fully adjust to a change in radiative forcing so that only initial and final states are known, not the time dependent response. Alternatively, a prescribed emission scenario can be explored (e.g., percent increase of CO_2 per year) and the nonequilibrium time-dependant response is analyzed.

3.2 Model types

Four commonly used types of climate models are briefly described below. Atmosphere-Ocean-General Circulation Models (AOGCMs), used at various resolutions, were the standard model used in the fourth IPCC report (IPCC-AR4, 2007) and they continue to be used extensively today. This type of model focuses on the physical components of the climate system (ocean, land, sea ice, atmosphere) and is used to make projections based on GHGs and aerosol forcings. This model is also used for regional studies (sub-continental scale) and when biogeochemical feedbacks are not critical. These models may have up to 50,000 horizontal grid points.

Earth system models (ESMs) are currently the state-of-the-art model and they include biogeochemical feedbacks (e.g., carbon cycle; sulfur cycles and ozone). They may have up to 100,000 horizontal grid points. Earth system models of intermediate complexity (EMICs) have lower resolution than ESMs and more idealized components of the Earth system. They are used for certain questions such as feedbacks on a millennial time scale or where a large number of ensembles is required. They may include ice sheets among their components, which are not always included in ESMs; they are efficient and flexible models. Regional Climate Models (RCMs) can be used for regional (sub-continental) studies in specific geographic areas as downscaled global models.

3.3 Model improvements

Improving climate models is an iterative process involving additional model components, improved understanding of the processes involved in the climate system and, as more powerful computers become available, increased spatial and temporal resolution. Improvements to various aspects of climate system models since IPCC-AR4 (2007) include land use/cover change, ocean-atmosphere interactions, sea ice properties, and biogeochemical cycles.

In the case of the land, factors, such as soil type, soil moisture, vegetation, wildfires, snow cover and groundwater all play a role in climate change, commonly through changes in surface albedo, and current models take into account several of these variables. Models should also take into account heat and water fluxes between the atmosphere and the ground. Since IPCC-AR4 (2007), newer models include vegetation dynamics and CO_2 exchange with the atmosphere. In the case of the atmosphere a wide range of processes are included in newer models: convection, cloud-microphysical and aerosol processes, and aerosol-radiation interactions and ground-lower atmosphere interaction (see Chapter 2).

The basic thermodynamics of sea ice is well understood. Different albedo values are attributed to bare ice, melting ice, snow covered ice, and open water. Newer models include solar-radiation ice interactions and ice thickness and ice deformation processes as well as inclusion of salinity, chemistry, and biogeochemical interactions. Ice microstructure is also receiving increased attention.

A limitation of AOGCMs is the lack of feedback between physical, chemical, and biogeochemical interactions and ESM models are designed to rectify this situation by including biogeochemical cycles in the climate system. Given that the two most important GHGs are CO_2 and CH_4, this inclusion is an important advance. Some of these models include both the terrestrial and oceanic carbon cycles. Ocean uptake of CO_2 is highly variable both spatially and temporally and depends on several variables including the composition of the marine ecosystem (e.g., plankton groups). Newer models also include the sulfur and nitrogen cycles as well as ozone chemistry. Models can also include ocean acidification, which is a negative feedback on atmospheric CO_2 (Ridgwell et al., 2007).

As mentioned earlier, aerosols include black carbon (soot), dust, sea salt, sulfate (from oceanic dimethylsulfide) and nitrates. Climate models (AOGCMs and ESMs) that include physical representations of aerosols often improve the historical and present-day climate simulations. Aerosol-cloud interaction can include complex effects involving cloud droplet size and number (see Chapter 9). Some ESMs and EMICs include the methane cycle and emissions from wetlands as the climate warms. Other models include the carbon stock in permafrost as it thaws. Global vegetation and wildfires have important implications for the terrestrial carbon sink. The northward migration of boreal forest is an example of a biogeophysical feedback on the climate system. Wildfires, which become more widespread with increasing temperature, are a source of soot (black carbon) aerosols that affect the climate system in terms of radiative forcings. Some ESMs include land ice sheets; however, the magnitude and seasonality of melt-water from Greenland and Antarctica ice sheets is a major source of uncertainty (Sutherland et al., 2019).

Resolution in AOGCMs and ESMs is typically one or two degrees (latitude/ longitude) for the atmosphere and one degree for the ocean. Vertical layers for the atmosphere number 30 to 40 and in some cases as high as 80 and for the ocean, typically 30 to 60 layers. The resolution for regional climate models (RCMs) can be as high as 10 km.

3.4 Model evaluation

The essential approach to model evaluation is to compare simulated quantities (e.g., global temperature, precipitation, etc.) to observational measurements. Evaluation of model errors involves isolating individual processes in a model (e.g., cloud regimes, ocean circulation) in addition to isolating individual model parameters. Satellite data cover a wide variety of meteorological conditions, and they are also global in scope and play an important role in model evaluation.

Weather forecasting uses the present state of the climate as its primary input. In contrast, climate models use in advance the statistical mean of weather on seasonal to century timescales. The atmosphere component of climate models can also be used in weather forecasting. It was learned that these climate models could go astray after only a few days, highlighting the importance of rapid sub-grid scale processes without longer-term feedbacks. This learning experience now allows testing of tuning parameters for these sub-grid scale processes in the atmosphere for climate simulation.

3.5 Ensemble approach to evaluation

There are two types of ensembles: multimodel ensembles and perturbed parameter ensembles. The first type is created from multiple climate research centers and usually has a small sample size and, since some models share the same components, they are not all independent of each other. Commonly, the results are presented simply as the arithmetic mean of the ensemble with each member having the same weight (e.g., Kirtman et al., 2013). The perturbed parameter ensembles assess the uncertainty of a single model by varying the input parameters. Statistical methods are used to estimate which parameters drive most of the uncertainty. No single model evaluation method is used but rather the combined use of several evaluation methods is best. An obvious criterion for any climate model used to make climate projections is that it must be able to represent historical and current climate.

3.6 Model intercomparisons

Organized intercomparison projects (e.g., CMIP3 and CMIP5) provide, via benchmark experiments, tests that allow a model's ability to simulate observed historical and paleoclimates and also projections decades into the future. When researchers perform a common experiment, the results can be compared to each other and to observed climate. This allows the strength and weaknesses of different models to be evaluated in a controlled setting. Coupled Model Intercomparison Project phase 3 (CMIP3) was assessed in IPCC-AR4 (2007) and a more comprehensive project (CMIP5) included specification of radiative forcing using both AOGCMs and ESMs. Most model simulations reported in IPCC-AR5 (2013) are historical (1850–2005)

Table 3.1 Mean temperature increase and sea level rise twenty-first century.

RCPs	Temp. (°C) 2046–2065	Temp (°C) (2018–2100)
RCP 2.6	1.0	1.0
RCP 4.5	1.4	1.8
RCP 6.0	1.3	2.2
RCP 8.5	2.0	3.7
	Sea level rise (m)	Sea level rise (m)
RCP 2.6	0.24	0.4
RCP 4.5	0.26	0.47
RCP 6.0	0.25	0.48
RCP 8.5	0.30	0.63

Source with permission: adapted from Collins et al., 2013, (SPM.2).

but also include simulations of the mid–Holocene (circa 6 thousand years ago) and also the last glacial maximum (circa 21 thousand years ago). Simulations involving a percentage increase in CO_2 per year are used to evaluate the transient response to climate forcings; equilibrium climate responses use a fourfold or twofold increase in CO_2.

In the case of CMIP5, prescribed input included anthropogenic emissions, land use change and a prescribed set of radiative forcings (see Table 3.1). How to deal with emissions of natural aerosols (dust, sea salt, volcanic ash) were left up to the individual research groups. For AOGCMs forcing agents were a prescribed set of GHG concentrations including halocarbons and other species. Land change use included partitioning among pasture, cropland, forest and wood harvest, and urban land. Sunspot activity was also specified for the period 1850–2008. As noted already, ESMs included biogeochemical cycle feedbacks.

3.7 Results

3.7.1 Temperature and precipitation

The mean air temperature (at 2 m elevation) based on the mean of available CMIP5 models agree with observations within 2°C for the period 1980–2005. Areas with larger errors include high elevations (e.g., Himalayas), Greenland, and Antarctica ice sheets and over ocean upwelling regions (e.g., South America and Africa) (Fig. 3.1A). Models also simulate seasonal differences (means for December-January-February versus June-July-August). Seasonal differences over the oceans are small compared to those over the continents due to higher heat capacity of ocean water and slower advection

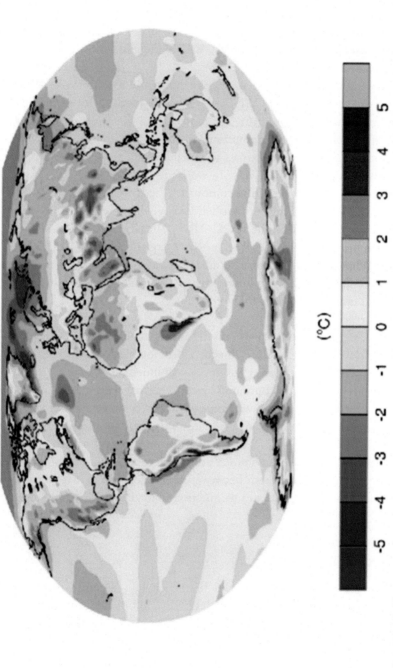

Figure 3.1A The error in the annual mean surface air temperature for CMIP5 models relative to observations. In most areas the multi-model mean agrees within 2°C of observations. But there are several areas where the error is larger namely, over high elevations, Greenland and Antarctica and over ocean upwelling regions. (*Source with permission: Flato et al., 2013.*)

(°C)

-5   -4   -3   -2   -1   0   1   2   3   4   5

processes. The models, however, tend to produce larger seasonality than observed over land areas and underestimate seasonality over the oceans. Model simulation of precipitation is more difficult than temperature and simulations are less accurate but reproduce the major global trends (e.g., higher precipitation at higher latitudes), with the exception of the tropics where significant errors occur (Fig. 3.1B).

The mean of CMIP5 models for the historical record (1870–2010) relative to the reference period (1961–1990) do a good job of simulating surface temperature with departures of less than 0.5°C from observations. Cooling after major volcanic eruptions are also well simulated and the acceleration in warming over the past several decades is simulated (Fig. 3.2). The exception to this is the "hiatus" in warming 10 to 15 years ago where models continue to warm but observations reached a plateau. This can be seen in Fig. 3.2 where the mean of models is above the observational record over the period 2000–2010. The IPCC-AR5 (2013), Box 9.2, discusses the cause of this hiatus and provides three possible explanations:

- Internal climate variability on a decadal scale, noting that CMIP5 models have a predictive range on a longer than 20 year timescale.
- Incorrect effective radiative forcing (ERF).
- Model error.

The NCA4 (2017, Box 1.1) also discusses this same hiatus and suggests that deeper levels in the ocean absorbed heat over this period leading to cooler surface temperatures. Sea level continued to rise during this period indicating the ocean continued to store thermal energy. More recent observational data indicate the warming trend has resumed following the hiatus, in agreement with models. The hiatus has important implications for climate projections as the temperature plateau may be used as part of the initial conditions in some model projections. This topic is still the subject of ongoing research. The mismatch between observation and climate models has been used by climate skeptics to question climate models (e.g., Singer and Avery, 2008).

3.8 The ocean

The ocean plays several crucial roles in the climate response including transient uptake of CO_2 and heat, sea level rise, and climates modes such as El Niño. Over the depth range 200 m to 2000 m and over most latitudes the CMIP5 multimodel mean ocean temperature is between 1°C and 3°C too warm and at deeper levels is about 1°C too cold. In addition, above

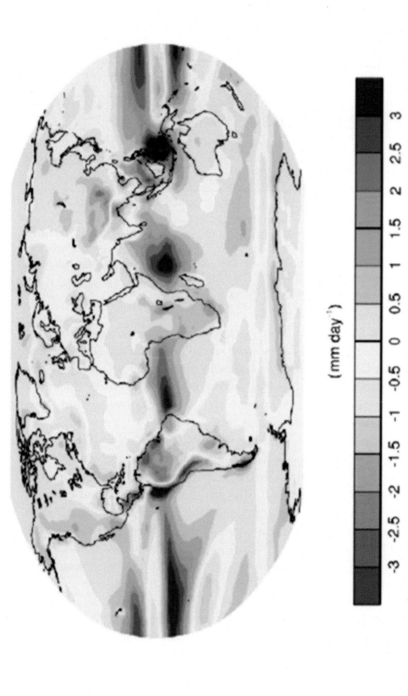

(mm day^{-1})

Figure 3.1B The difference between multimodel mean (CMIP5) and observations for annual mean precipitation rate (mm per day). The broad scale of features of precipitation are in modest agreement with observation, but there is significant disagreement over the tropics where precipitation is too high. (*Source with permission: Flato et al., 2013.*)

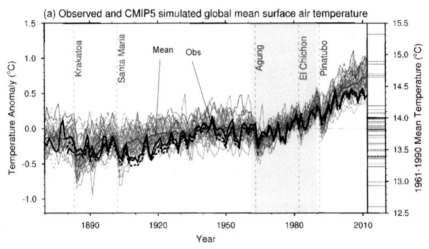

Figure 3.2 Observed and annual mean global surface temperature anomalies relative to the period 1961–1990 for EMIC models (thin lines). Vertical lines are major volcanic eruptions which are followed by cooling. The vertical shaded region is the reference period. *(Source with permission: Flato et al., 2013.)*

the 200 m level it is too cold by about 1°C near the surface. The Array for Realtime Geostrophic Oceanography (Argo) provides good coverage for salinity observations. The historical record for sea surface salinity is well simulated in CMIP3 models, but regionally, biases of up to 2.5% (parts per thousand) occur. As noted earlier, salinity is a function of the difference between precipitation and evaporation. The model biases appear to be related to surface freshwater influxes (Flato et al., 2013).

How fast are the oceans warming? The IPCC-AR5 report (Rhein et al., 2013) presented five estimates for the upper ocean heat content (UOHC) within the 0-700 m depth range over the time period 1950 to 2010. The data are highly variable prior to 1970 due to gaps in spatial and temporal coverage. From 1971 to 2010 the data still show substantial uncertainty with mean values of UOHC for the five studies showing a wide range: 137, 118, 108, 98, 74 TW (1 TW = 10^{12} watts) for a mean value of 107 ± 20 TW (note large error). For an ocean area of 363 × 10^6 km^2 this mean value converts to a warming rate of 0.29 ± 0.05 Wm^{-2}. These studies corrected the instrumental data in different ways when extrapolating to areas with data gaps resulting in substantially differences in the results. Moreover, the results are lower than the output from climate models in CMIP5 models. The average of these models was 0.39 ± 0.07 Wm^{-2} for the period 1971–2010 (Cheng et al., 2019).

Several more recent studies have attempted to account for spatial and temporal gaps in ocean temperature measurements using different methodologies (Cheng et al., 2017; Ishii et al., 2017; Resplandy et al., 2018). Domingues et al. (2008) used satellite altimetry to complement sparse areas of coverage (satellite altimetry can be used to estimate the proportion of sea level rise due to thermal expansion). These newer studies show highly consistent results, higher than the IPCC-AR5 results. Three of the recent studies listed above show warming rates of 0.36 ± 0.05, 0.37 ± 0.04, and 0.39 ± 0.09 Wm^{-2}. These results are similar to the CMIP5 model outputs (0.39 ± 0.07 Wm^{-2}).

The deployment of Argo floats globally has resulted in superior coverage and reduced uncertainties since the 2000s. Observations over the period 2005–2017 suggest an accelerated warming trend of 0.54 ± 0.02, 0.64 ± 0.02, and 0.68 ± 0.06 Wm^{-2}. The mean of CMIP5 results for the same time period is 0.68 ± 0.02 (Cheng et al., 2019) showing good agreement between observations and models.

The UHOC estimates in the IPCC-AR5 (2013) are now recognized to have been too low and more recent estimates, corrected for instrumental bias, are higher and agree with computer climate models. More contemporary estimates based on Argo results suggest an acceleration in ocean warming since 2005. Cheng et al. (2019) use CMIP5 models to predict ocean heat content out to 2100 under different Representative Climate Pathways (RCP8.5 and RCP2.6; see Table 3.1). This has consequences for sea level rise due to thermal expansion (not including ice melting) and intensification of the hydrological cycle reflected in storms, hurricanes, and extreme precipitation events (Box 3.1).

BOX 3.1 Modeling paleoclimates

This case study is based on Hargreaves et al. (2013) who studied the Last Glacial Maximum (LGM) and the mid-Holocene climate. The LGM occurred about 21000 years BP (before present) and the mid-Holocene dates to 6000 years BP. An attractive aspect of testing the performance of models on simulating paleoclimates is that the models were not developed based on these climates and therefore provide an independent test of the models under different climate conditions. The LGM involved the presence of large northern hemisphere ice sheets, altering the topography, and had low CO_2 concentrations (~185 ppm) compared to pre-industrial times (~280 ppm). Only small changes in orbital forcing were involved.

(continued)

Box 3.1 Modeling paleoclimates— *cont'd*

The mid-Holocene, on the other hand, had a much milder climate characterized by higher GHGs compared to the LGM (methane concentration of 650 ppb, lower than the pre-industrial value of 750 ppb); it had significantly different seasonal orbital forcing in the northern hemisphere.

The climate models included both AOGCMs and AOGVCMs, where V indicates vegetation feedbacks were included (Hargreaves et al., 2013). The models required input data for sea surface temperature and surface air temperature. The proxy databases for the LGM are extensive and include both the oceans (deep sea cores) and the continents (tree pollen), whereas for the mid-Holocene, ocean data are relatively sparse and on the continents the data were clustered in Europe and North America. In addition to the problems with data coverage for the mid-Holocene, this period lacks the strong mean annual orbital forcing present at the LGM, making simulation of the Holocene climate more of a challenge.

In the case of the LGM the model results generally do a good job of modeling the magnitude of the cooling, being slightly better over the oceans than the continents. The land-ocean difference in temperature and polar amplification were also simulated appropriately. In the case of the mid-Holocene the models did a poor job of modeling the seasonality on land. Modeling of seasonality over the oceans was not attempted (due to lack of coverage). It was suggested that vegetation feedbacks and aerosol feedbacks may be partly to blame for the poor performance of the models in the Holocene (Hargreaves et al., 2013; Flato et al., 2013).

3.9 Carbon cycle

Since IPCC-AR4 (2007) the transition from Atmosphere-Ocean-General Circulation Models (AOGCMs) to Earth System Models (ESMs) was prompted by the recognition of the importance of the carbon cycle in the simulation of twenty-first century climate (Flato et al., 2013). Since that time many ESMs have incorporated interactive carbon cycles. These CMIP5 models use prescribed RCP pathways (see Table 3.1) to calculate the atmosphere-ocean and atmosphere-land CO_2 fluxes. The biggest uncertainty is on the land due to limited direct observations on things such as soil carbon and vegetation carbon on sufficiently large scales. Nevertheless, two thirds of CMIP5 models are within 50% of direct observations. The ensemble mean of 23 CMIP5 models do a good job of simulating the direct observations of the Global Carbon Project (Le Quere et al., 2009). In the case of the atmosphere-ocean flux, the Global Carbon Project however overestimates the mean value for the period 1960–2005 by about 20%

compared to the simulation models. Whether the observations or the models are at fault is unclear. The mean model ensemble is nevertheless within one standard deviation of the observations. Before looking at near-term and long-term projections, the Paris Accords and Representative Climate Pathways are briefly discussed.

3.10 The Paris Accords

The Paris agreement on climate change went into effect on December 15, 2015. It is a relatively short document and it has 27 articles over 25 pages. About 190 parties (countries), including the European Union, signed the agreement including the United States and China, the latter two countries representing more than 40% of global emissions. As soon as President Trump was elected in 2016 he said the United States would withdraw from the agreement – but this withdrawal can only go into effect after 4 years from the time of withdrawal. Several states, especially California, are attempting to live up to the terms of the Accord. A recurrent theme throughout the agreement is that developed nations should help underdeveloped nations in the transition to lower emissions through technological transfer and financial aid.

The most important and notable feature of the accord is that reduction of emissions are up to each individual nation to specify themselves and are voluntary. The major goal of the accord is to limit global warming by the end of the century to less than 2°C over preindustrial levels. Since we already have 1°C warming since pre-industrial time, this means we must limit additional warming to less than 1°C by century's end. Additional recurring themes in the accord are that the transition to low emissions should be sustainable and should protect indigenous peoples and other vulnerable populations (e.g., inhabitants of small low lying islands) with an emphasis on social justice, reducing poverty, and also protecting ecosystems. Nations must periodically report to the United Nations on how they are progressing on their emission pledges. Note: President Biden re-entered the Paris Accords after his election in 2020.

3.11 Representative climate pathways

The first IPCC-AR report (1990) introduced four emission scenarios as future possibilities as a result of climate change and they represent a range of plausible pathways or targets. They were based on population and economic growth projections of developed and developing nations and on the possible effects of technological change and energy efficiency. Emissions of the greenhouse gases were estimated out to the end of the century

(2100). The four scenarios were: business as usual (or scenario A) being the highest emissions pathway and scenarios B, C, and D with decreasing emissions. In scenario A, fossil fuel use is intensive (mainly coal) and tropical forests are depleted. In scenario B, a switch to natural gas takes place and energy efficiency increases. In scenario C, nuclear energy and renewables are mainstays and CFCs are no longer produced. Lastly, in scenario D, CO_2 emissions are reduced by 50% compared to 1985 levels. These scenarios included fixed targets for greenhouse gas emissions as well as transient increases. Subsequent IPCC reports of likely emission scenarios were more comprehensive and included updated population projections and economic forecasts and technical advances.

Representative climate pathways (RCPs) were introduced briefly in Chapter 2 and they represent future radiative forcing scenarios (RCP2.6, RCP4.5, RCP6.0, and RCP8.5) where the numbers represent radiative forcings in W/m^2. As noted earlier these forcings are measured at the top of the troposphere, and the stratosphere is allowed to come to equilibrium while the troposphere temperature is kept constant. They were introduced for the first time in IPCC-AR5 (2013) and NCA4 (2017). They include land use and cover changes, all the GHGs and water vapor as well as ozone and aerosols. From Table 3.1 it can be seen they are associated with concentration ranges and emissions of CO_2. Table 3.1 shows the temperatures associated with the four pathways for two time periods and also the sea level rise expected for each scenario. Representative Climate Pathways can be paired with Shared Socioeconomic Pathways (SSPs) which allow mitigation and adaptation responses by various global societies and communities to be evaluated depending on their socioeconomic status and other factors, such as economic development and government style (see Chapter 7).

3.12 Near-term climate projections

Near-term projection refers to simulations out to the mid–century (2050) whereas long-term projections extend to end of the century or longer (IPCC-AR5, 2013, Chapters 11 and 12 in that publication). Near-term projections are more sensitive to initial conditions and less sensitive to long term forcings (i.e. different RCPs) and the reverse is true for long-term projections where initial conditions are not very important but are more sensitive to different RCPs.

A word on some terminology is in order before proceeding. The innate behavior of the climate system (i.e., chaotic, nonlinear) imposes limits on the ability to predict its evolution. Two types of variability are recognized: internal and external. Internal variability refers to the natural variability of the climate system (specifically, it might refer to a preindustrial Earth where the climate is at equilibrium and where solar changes are absent). External variability refers to various anthropogenic and natural forcings (e.g., GHGs or volcanic eruptions discussed in Chapter 2). In the case of temperature (T), this can be written as $T(t) = T_f(t) + T_i(t)$ where t is time and T_i is the internal forcing and T_f is the external forcing. The predictability of a model to simulate climate is considered low if small changes in initial input result in diverging output (sometimes referred to as the butterfly effect). It is a feature of the physical system itself, not the ability of a model to make a good forecast. Other terms used by climate scientists which have specific meaning are: accuracy, skill, and reliability. Accuracy is the error between a model (forecast or hindcast) and observation and it was discussed above with regard to global temperature and precipitation (Fig. 3.1). Skill measures the quality of a forecast relative to some benchmark or reference forecast. The skill score is defined as $S = 1-E_f/E_{ref}$ where E_f is the error on the forecast and E_{ref} is the error on the reference model. Here error refers to any commonly used statistical test (e.g. standard deviation or root mean squared). If the forecast has no errors, it has a skill score of one. If it has the same number of errors as the reference forecast its skill level is zero. If the model has more errors than the reference model a negative score results (Hargreaves at al., 2013). Reliability refers to how well the predicted probability distribution matches the observed frequency of the forecast event.

In IPCC-AR5 (2013) near-term projections extend out to 2050 and they include historical simulations based on industrial era conditions to the present, so that historical data serve to initialize the models. Simulations into the future are based on CMIP5 models using representative pathway scenarios (RCPs) as forcings. However, because near-term projections are not very sensitive to RCPs, near-term projections in IPCC-AR5 (2013) were reported largely using a single intermediate forcing, namely RCP 4.5.

An important question for climate model projections is what the externally forced signal of near-term climate change is and how large it is compared to natural internal variability? In this regard, three sources of uncertainty are identified in climate models (Fig. 3.3):

- The natural internal variability of the climate system (e.g., storm tracks and climate modes, such as El Niño). This variability places limits on the precision with which future climate can be projected.

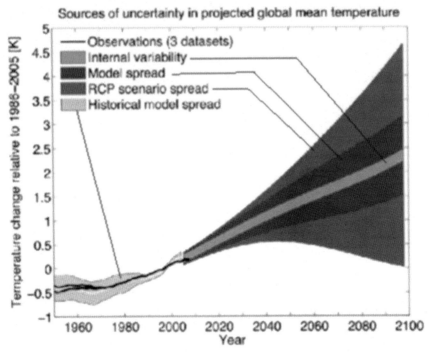

Figure 3.3 Sources of uncertainty in climate projection out to 2100 as a function of time for CMIP5 results. Internal variability is shown by the narrow innermost band. Model spread is shown by the next band and the outer band is based on RCP scenario spread. Historical model spread is also shown. Internal variability is more important for near-term projections and RCP scenarios become more important for long-term projections. *(Source with permission: Kirtman et al., 2013.)*

- Uncertainty regarding past, present, and future natural and anthropogenic forcing. Some constraints on the first two variables are provided by the historical record and current observations. Future forcings, on the other hand, are much more challenging to predict.

- Uncertainty over the *response* of the climate to specific forcings. In this regard, chapter 15 in NCA4 (Kopp et al., 2017) discusses potential surprises in climate response and tipping points.

The signal to noise ratio is an important quantity in climate modeling, where the signal is the magnitude of the climate change. The noise can be regarded as the total uncertainty due to the three factors outlined above. The farther out the projection is time wise, the lower the signal to noise ratio.

Fig. 3.4 shows CMIP5 projections of global mean temperature under the RCP 4.5 scenario for the period 2016–2050 relative to the reference period 1986–2005. The range for the projected temperature anomaly in 2050 is 0.47°C to 1.0°C (5% to 95% range of models). Given that the current global

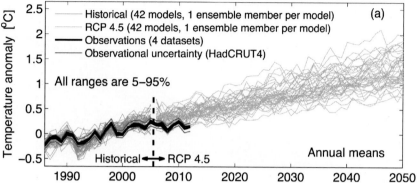

Figure 3.4 Global annual mean surface air temperature anomaly relative to the period 1986–2005 for the RCP4.5 scenario from CIPM5 models (thin lines). Historical observation is shown by the thick black line. For the period 2016–2035 the temperature anomaly is 0.47°C to 1.0°C. *(Source with permission: Kirtman et al., 2013.)*

temperature rise since preindustrial times is $0.8 \pm 0.2°C$ (Chapter 1), the temperature rise in 2050 since pre-industrial times would be approximately 1.5°C to 2.0°C. This RCP pathway would therefore satisfy the Paris Accords out to 2050. The special IPCC (2018) report examined the different climate impacts between these two temperature rises (1.5°C and 2.0°C) and found surprisingly large differences (the results of this report are summarized in Chapter 8). The RCP4.5 pathway was chosen because, of the four pathways (Table 3.1), it is intermediate in its radiative forcing, and it should be borne in mind that the real climate may follow a path above or below these estimates. In addition, there may be processes in the real climate that are not adequately simulated by the CMIP5 models. Without a crystal ball, science can only make reasonable assumptions for projections.

3.13 Long-term projections

Projections of climate change are uncertain for three reasons:
- Future natural and anthropogenic forcings are uncertain
- Incomplete understanding and imprecise models of the climate system
- Existence of internal climate variability

The Climate Model Intercomparison Project phase 5 (CMIP5) together with the RCP scenarios outlined above play an important role in long-term projection out to the end of the century and longer (IPCC-AR5, 2013; NCA4, 2017). As mentioned already, in the case of long-term projection, the RCP path followed plays a more important role than the initial conditions.

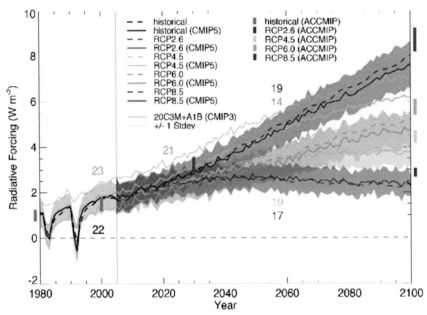

Figure 3.5 The global mean radiative forcing for the period 1980–2100 by three methods: RCP scenarios (dashed lines), CMIP5 models using RCP scenarios (solid lines) and results from the Atmospheric Chemistry and Climate Model Intercomparison Project (ACCMIP; shaded regions, 68% confidence interval) using the RCP scenarios. The number of models for CMIP3 and CMIP5 is shown. Historical data for the period 1980 to 2005 is also shown. The three methods are in reasonable agreement. *(Source with permission: Collins et al., 2013.)*

Many CMIP5 models use Earth System Models (ESMs) and they can use either GHG concentrations or GHG emissions as input and these quantities can be calculated from each other. A common reference baseline in these long-term projections is the historical record for the period 1986–2005.

The global mean radiative forcing for the twenty-first century is an important quantity as it reflects the global energy balance and the resultant climate change. Fig. 3.5 shows this quantity over the period 1980–2100 based on three related estimates. The dashed lines show the four RCP scenarios and the solid lines show the results from CMIP5 models using these same RCPs. The shaded regions are the results from the Atmospheric Chemistry Climate Model Intercomparison Project (ACCMIP). The agreement between all three estimates is good. An important conclusion of these studies is that variation of natural forcings during the industrial era is a small fraction of the anthropogenic forcings for the twenty-first century. Table 3.2 shows the temperature anomalies (relative to the reference period

Table 3.2 CMIP5 annual mean global surface air temperature anomalies (°C)[a]

Time period	RCP 2.6	RCP 4.5	RCP 6.0	RCP 8.5
2046–2065	1.0 ± 0.3	1.4 ± 0.3	1.3 ± 0.3	2.0 ± 0.4
2081–2100	1.0 ± 0.4	1.8 ± 0.5	2.2 ± 0.5	3.7 ± 0.7

[a] Reference period 1986–2005; error one sigma; *Source:* Adapted from IPCC-AR5, WGI, Table 12.2.

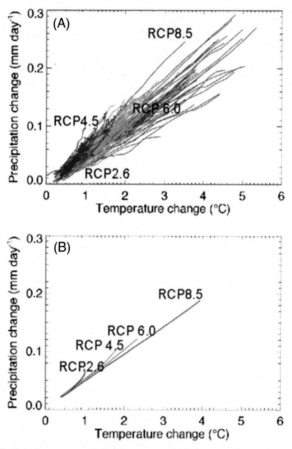

Figure 3.6 (A) Global mean precipitation (mm per day) versus temperature (°C) change relative to 1986–2005 for multimodel CMIP5 (one line per model) for the four RCPs projected out to 2100. Note the linear relationship between temperature and precipitation. (B) Corresponding multi-model means for each RCP. *(Source with permission: Collins et al., 2013.)*

1986–2005) associated with RCPs for two different time periods. Note that the temperature anomalies are somewhat similar for the earlier time period but for the end century scenario major temperature anomalies occur. This would suggest that policymakers should focus on the 2050 timeframe to limit emissions; otherwise things may get out of control. To convert these results to a reference frame of 1850–1900, 0.61°C should be added.

Long-term projections indicate that both mean global temperature and precipitation will increase linearly together over this century (Fig. 3.6). In the case of precipitation, both a fast and a slower longer-term response is recognized. An increase in CO_2 decreases the radiative cooling of the troposphere and reduces evaporation and rainfall on a short timescale (Andrews et al., 2010). On a longer time scale, a slow increase in temperature increases atmospheric moisture and hence precipitation (Fig. 3.6). On a regional scale (subcontinental) models suggest that wet areas will become wetter and dry areas become drier (Boucher et al., 2013).

References

Andrews, T., Forster, P., Boucher, O., Bellouin, N., Jones, A., 2010. Precipitation, radiative forcing and global temperature change. Geophys. Res. Lett. 37, L14701.

Boucher, O., Randall, D., Artaxo, P., Bretherton, C., Feingold, G., others, 2013. Clouds and Aerosols. In: Stocker, T.F., Qin, D., Plattner, G.K., Tignor, M., Allen, S.K., Boschung, J., Nauels, A., Xia, Y., Bex, V., Midgley, P.M. (Eds.), Climate Change 2013: The Physical Basis. Contribution of Working Group 1 to the Fifth Assessment of the Intergovernmental Panel on Climate Change. Cambridge University Press, UK and New York.

Cheng, L., Abraham, J., Hausfather, Z., Trenberth, K.E., 2019. How fast are the oceans warming? Science 363, 128–129.

Cheng, L., Trenberth, K.E., Fasullo, J., Boyer, T., Abraham, J., Zhu, J., 2017. Improved estimates of ocean heat content from 1960 to 2015. Sci. Adv. 3, e1601545.

Collins, M., Knutti, R., Arblaster, J., Dufresne, J-L., Fichefet, T., Friedlingstein, P., eight others, 2013. Long-term Climate Change: Projections, Commitments and Irreversibility. In: Stocker, T.F., Qin, D., Plattner, G.K., Tignor, M., Allen, S.K., Boschung, J., Nauels, A., Xia, Y., Bex, V., Midgley, P.M. (Eds.), Climate Change 2013: The Physical Basis. Contribution of Working Group 1 to the Fifth Assessment of the Intergovernmental Panel on Climate Change. Cambridge University Press, UK and New York.

Domingues, C.M., Church, J.A., White, N.J., Gleckler, P.J., Wijffels, S.E., Barker, P.M., Dunn, J.R, 2008. Improved estimates of upper-ocean warming and multi-decadal sea-level rise. Nature 453, 1090–1093.

Flato, G., Marotzke, J., Abiodun, B., Braconnot, P., Chou, S.C., Collins, W., Cox, P., et al., 2013. Evaluation of climate models. In: Stocker, T.F., Qin, D., Plattner, G.K., Tignor, M., Allen, S.K., Boschung, J., Nauels, A., Xia, Y., Bex, V., Midgley, P.M. (Eds.), Climate Change: The Physical Science Basis. Working Group I Contribution to the Fifth Assessment Report of the Intergovernmental Panel on Climate Change. Cambridge University Press, UK and New York.

Hargreaves, J.C., Annan, J.D., Ohgaito, R., Paul, A., Abe-Ouchi, A., 2013. Skill and reliability of climate model ensembles at the last glacial maximum and the mid-holocene. Climate Past 9, 811–823.

IPCC – AR5, 2013. Climate change 2013. The physical science basis. In: Stocker, T.F., Qin, D., Plattner, G.K., Tignor, M., Allen, S.K., Boschung, J., Nauels, A., Xia, Y., Bex, V., Midgley, P.M. (Eds.), Working Group I Contribution to the Fifth Assessment Report of the Intergovernmental Panel on Climate Change. Cambridge University Press, UK and New York.

IPCC 2018. Global warming of 1.5°C an Intergovernmental Panel on Climate Change special report of global warming of 1.5°C above pre-industrial levels. In: Masson-Delmotte, V., Zhai, P., Pörtner, H. O., Roberts, P., Skea, J., Shukea, P. R., Pirani, A., Moufouma-Okia, A., Péan, C., Pidcock, R., Connors, S., Matthews, J. B. R., Chen, Y., Zhou, Y., Gomis, M. I., Lonnoy, E., Maycock, T., Tignor, M. and Waterfield, T. (eds.) Cambridge University Press, UK and New York.

IPCC- AR4, Climate change, 2007. The physical science basis. In: Solomon, S., Qin, D., Manning, M., Marquis, M., Averyt, K., Tignor, M.M.B., Miller, H.L., Chen, Z. (Eds.), Contributions of Working Group I to the fourth Assessment Report of the Intergovernmental Panel on Climate Change. Cambridge University Press, UK and New York.

Ishii, M., Fukuda, Y., Hirahara, S., Yasui, S., Suzuki, T., sato, K., 2017. Accuracy of global upper ocean heat content estimation expected from present observational data sets. SOLA 13, 163–167. doi:10.2151/sola.2017-030.

Kirtman, B., Power, S.B., Boer, G.J., Bojariu, R., Camilloni, I., et al., 2013. Near-term climate change: projections and predictability. In: Stocker, T.F., Qin, D., Plattner, G.K., Tignor, M., Allen, S.K., Boschung, J., Nauels, A., Xia, Y., Bex, V., Midgley, P.M. (Eds.), Climate Change 2013: The Physical Basis. Contribution of Working Group 1 to the Fifth Assessment of the Intergovernmental Panel on Climate Change. Cambridge University Press, UK and New York.

Kopp, R.E., Hayhoe, K., Easterling, D.R., Hall, T., Horton, R., Kunkel, K.E., LeGrande, A.N., 2017. Potential surprises - compound extremes and tipping elements. In: Wuebbles, D.J., Fahey, D.W., Hibbard, K.A., Dokken, D.J., Stewart, B.C., Maycock, T.K. (Eds.). Climate Science Special Report; Fourth National Climate Assessment, 1. U.S. Global Change Research Program, Washington D. C., USA.

Le Quere, C., Raupach, M.R., Woodward, F.I, 2009. Trends in sources and sinks of carbon dioxide. Nature Geosci 2, 831–836.

NCA4, 2017. Climate Science Special Report. In: Wuebbles, D.J., Fahey, D.W., Hibbard, K.A., Dokken, D.J., Stewart, B.C., Maycock, T.K. (Eds.). The United States Government's Fourth National Climate Assessment, 1. U.S. Global Change Research Program, Washington D. C., USA.

Resplandy, L., Keeling, R.F., Eddebbar, Y., Brooks, M.K., Wang, R., Bopp, L., Long, M.C., Dunne, J.P., Koeve, W., Oschlies, A., 2018. Quantification of ocean heat uptake from changes in atmospheric O_2 and CO_2 composition. Nature 563, 105–108.

Rhein, M., Rintoul, S.R., Aoki, S., Campos, E., Chambers, D., Feely, R.A., Gulev, S., Johnson, S.A., Kostianoy, A., Mauritzen, C., Roemmich, D., Talley, L.D., Wang, F., 2013. Observations: Ocean Climate change. In: Stocker, T.F., Qin, D., Plattner, G.K., Tignor, M., Allen, S.K., Boschung, J., Nauels, A., Xia, Y., Bex, V., Midgley, P.M. (Eds.), The Physical Basis. Contribution of Working Group 1 to the Fifth Assessment of the Intergovernmental Panel on Climate Change. Cambridge University Press, UK and New York.

Ridgewell, A., Zandervan, I., Hargreaves, J.C., Bijma, J., Lenton, T.M., 2007. Assessing the potential long-term increase of fossil fuel CO_2 uptake due to CO_2-cacification feedback. Biogeosciences 4, 481–492.

Singer, S.F., Avery, D.T., 2008. Unstoppable Global Warming. Roman & Littlefield, New York.

Sutherland, D.A., Jackson, R.H., Kienholz, C., Amundson, J.M., Dryer, W.P., et al., 2019. Direct observations of submarine melt and subsurface geometry at a tidewater glacier. Science 365, 369–373.

CHAPTER 4

Paleoclimates

4.1 Introduction

Information from paleoclimate studies allows information on climate system processes (for example, feedbacks; Chapter 2), that are part of the preinstrumental record to be examined on longer timescales (centuries to millennia). Abrupt climate transitions on decade to century timeframes, such as major melting events and sea level changes, can also be examined (see Box 4.1). Such studies clearly have relevance to possible future climate projections. In addition, paleoclimate studies allow a determination of whether or not our current climate is unusual. For example, from studies of ice cores from Antarctica, we know that the mean global CO_2 concentration in the atmosphere (currently at 420 ppm in 2021) was never above 300 ppm in the last 800,000 years (Petit et al., 1999; Jensen et al., 2007). The current atmospheric composition therefore is unusual due to anthropogenic CO_2 emissions. We saw in Chapter 3 (Box 3.1) that computer modeling of paleoclimates (mid–Holocene and the Last Glacial Maximum) allows testing of climate models when applied to different climate situations for which the models were not originally designed. In addition, paleoclimate studies put constraints on the climate sensitivity parameter, λ, which relates radiative forcing to temperature (Chapter 2; Hegerl et al., 2006).

BOX 4.1 Sea-level rise at the Bølling warming.

An ancient well characterized sea level rise event serves as an example of major land-based ice melting at the end of the last glacial maximum at about 14.6 ka (Deschamps et al., 2012). Coral reefs are an important environmental indicator especially of sea level change. Within these calcium carbonate structures, multi-celled animals live and build their exoskeletons from calcium and oxygen to produce calcium carbonate ($CaCO_3$) from seawater. Coral reef animals are attuned to a variety of environmentally sensitive variables such as salinity, temperature (25°C typical of the tropics and sub-tropics) and abundant sunlight and oxygen (i.e., within the surf zone). They also do not tolerate detrital sediment input from adjacent rivers onshore – coral reefs usually show a break at these locations.

(continued)

Climate Change in the Anthropocene.
DOI: https://doi.org/10.1016/B978-0-12-820308-8.00010-6

Box 4.1 Sea-level rise at the Bølling warming.— *cont'd*

During "normal" times, such as the current Holocene epoch when sea level changes were small and gradual (with the exception of the 8.2 ka event), the corals could grow upward to keep pace with slowly rising sea level and so maintain the shallow water environment they require. Conversely, when sea level is falling, waves erode the coral, forming terraces, thereby still maintaining their shallow water environment. This makes coral reefs ideal markers in recording ancient sea level change and why geologists study coral terraces. By drilling into reefs in places like Barbados and Tahiti geologists can sample ancient coral reefs that used to be close to sea level but were drowned by the rapid rise in sea level as the ice melted at the end of the last glaciation. If the age of the corals and the depth below sea level of the terraces is known the rate of past sea level rise can be calculated, on the plausible assumption that the corals were always near sea level throughout.

Two methods of dating are used to find the age of the corals – one is the radiocarbon method (^{14}C) and the second less well known one is the uranium-thorium method (U-Th). In the radiocarbon method the amount of radioactive carbon-14 in the sample is measured and using the rate of decay of carbon-14 the age of the sample is calculated. In the thorium-uranium method, uranium decays to thorium and the age of the sample is calculated from the Th/U ratio when the rate of decay of uranium is known. Since the corals are composed of $CaCO_3$ and thorium does not easily fit into that lattice, they have no thorium initially when the coral forms so the measured Th/U ratio is directly related to the age of the coral.

The rise in sea level at the end of the LGM over the period of 16 ka to 12 ka years ago was about 60 m (~200 ft) (Fig. 4.4). On average the sea level rise was about 1 cm/yr (0.4 inches/yr) but during a particularly rapid sea level rise (14.3 ka to 14.65 ka) the annual rise was four times faster – this corresponds to what is called Melt Water Pulse 1A (MWP-1A) (Fig. 4.4). It was caused by melting of either the Laurentian (i.e., North American) or the Antarctic ice sheet or both (Deschamps et al., 2012). This rise in sea level would have caused the water table on land to rise to shallow levels and wetlands to expand globally. Ice cores from Greenland (Fig. 4.3) and Antarctica show a major rise in CH_4 at this time, the likely source being wetlands, especially in the tropics since northern latitudes would have been still frozen.

Many natural systems depend on climate and where such systems are preserved from the past it may be possible to derive paleoclimate information from them. By definition, proxy records of climate contain a climatic signal, but the signal may be weak and include extraneous factors (Bradley, 1999). Possibly the most important paleoclimate record is that preserved in ice cores (especially from Greenland and Antarctica) which provide information on GHG concentrations, the isotopic compositions (oxygen and hydrogen) of ice and air, dust concentration, and volcanic ash concentrations. Other important climate proxies include tree rings (dendrochronology), corals, tree pollen (palynology), cores from marine and fresh water sediments and glacial deposits in addition to cave deposits, to name a few (Bradley, 1999).

To convert a proxy record to quantitative information, the record must be calibrated and this is usually done using the instrument record, which extends back only about a hundred years in the case of temperature. It is commonly assumed that the amplitude of the proxy signal is linearly related to a climate variable (e.g., temperature or precipitation), but this may not be a good assumption. It does appear, however, that cumulative CO_2 emissions are linearly related to temperature (Chapter 2). Extrapolation of the calibrated record back thousands of years is fraught with uncertainties including gaps in the spatial and temporal record and errors in dating, the smearing out of abrupt transitions, and time lags of the proxy with climate change.

Different paleoclimatic records have different resolutions depending on the minimum sampling interval. High resolution records, such as tree rings, varved sediments, and ice cores are capable of annual resolutions, whereas deep ocean sediments may have a resolution of thousands of years, depending on the sedimentation rate; higher rates yield higher resolution and are less prone to bioturbation. Similarly, in the case of ice cores, regions of higher precipitation have higher resolution. In the case of tree pollen, since it takes centuries for forests to adjust to climate change, pollen records may have a resolution on this time scale and the signal maybe transitional rather than abrupt. In ice cores, there is a difference between the age of the ice and that of the trapped air bubbles, and this time lag increases with depth in the core and can be substantial (e.g., Petit et al., 1999); in addition, abrupt changes in climate maybe smeared out due to diffusion of the air trapped in the ice. This chapter first examines preindustrial external radiative forcings (e.g., orbital, solar, and volcanic), then addresses glacial and interglacial phases, followed by examination of climatic variations during selected periods of the Pleistocene and Holocene epochs.

4.2 Preindustrial external radiative forcings

4.2.1 Orbital forcing

Orbital forcing (also referred to as Milankovitch cycles after the Yugoslavian engineer) comprise three orbital variations that control the amount of solar radiation reaching the Earth over time (Fig. 4.1):

- Eccentricity – variations in the shape of the Earth's orbit with a periodicity of 100,000 years.

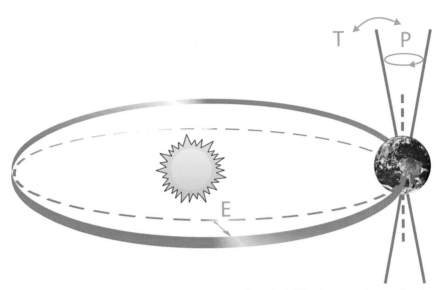

Figure 4.1 Earth's orbital changes (Milankovitch cycles) drive ice ages by varying incoming solar radiation. T denotes changes in the tilt (or obliquity) of the Earth's axis. E denotes changes in the eccentricity of the Earth's elliptical orbit around the Sun. P denotes precession, changes in the direction of tilt of the Earth's axis (or wobble). *(Source with permission: Jensen et al., 2007.)*

- Tilt of the Earth's axis (21.5° to 24.5°), also called obliquity, with a periodicity of 41,000 years.
- Precession (or wobble) with a periodicity of 23,000 and 19,000 years.

Milankovitch showed in the early twentieth century that when minima and maxima of these orbital variations coincide that solar radiation to the Earth reaches minimum and maximum values (Imbrie and Imbrie, 1986). The axial tilt controls the mean annual amount of radiation the Earth receives as a function of latitude. It was not until much later that it was shown from the study of oxygen isotopes in fossils from deep sea cores that these periodicities controlled the timing of glacial and interglacial periods (see below). More recently, it was also shown that these variations controlled the temperature, the sea ice extent, and glacier size during the last 2000 years (Jensen et al., 2007).

4.2.2 Solar forcing

We saw in Chapter 2 that solar irradiance is 1360 Wm^{-2} at the top of the atmosphere as measured by satellite observations. Over an eleven year solar cycle, solar irradiance can be separated into the total solar irradiance and the spectral solar irradiance in the ultraviolet part of the spectrum. The former exhibits changes of only about 0.1% and the latter changes by a few percent. Spectral irradiance largely affects the stratosphere whereas total solar irradiance affects the Earth's surface and these changes are due to magnetic phenomena on the sun's surface. Extension of these activities to pre-satellite era observations (~1978) is difficult but can be partly accomplished by sunspot counts back to about 1600 AD and also by records such as ice cores and tree rings and records of cosmogenic radionuclides such as ^{10}Be and ^{14}C.

4.2.3 Volcanic forcing

The cooling effects of major volcanic eruptions, which last only a few years, are due to production of sulfate aerosols. Computer models do a good job of simulating the cooling effects of these major eruptions, defined in IPCC publications as causing a negative radiative forcing of 1 Wm^{-2} or more (see Fig. 4.2). Radiative forcings are typically -1 to -5 Wm^{-2} but a few were as high as $-20Wm^{-2}$. For the interested reader some large historical eruptions were: Krakatoa, Indonesia (1889), Santa Maria, Guatamala (1902), Agung, Bali (1963), El Chichon, Mexico (1982), and Pinatubo, Philippines (1991).

Pre-satellite reconstructions use deposition of sulfate in ice cores from Greenland and Antarctica going back between 1500 and 1200 years (Jensen et al., 2007). Reconstructions are generally consistent with one another but do not always agree on the magnitude of aerosol injections and ice core chronologies also gives rise to some timing differences. Volcanic injections into the stratosphere versus the troposphere can be distinguished on the basis of sulfur isotopes in the cores. The recurrence time of large eruptions has a mean of 35 years (a range of 3 to 121 years) over the past thousand years.

4.3 High CO$_2$ worlds

In the Cenozoic era (65 Ma to the present) there is thought to have been three periods when the climate was warmer than present and pre-industrial CO_2 values were higher (Masson-Delmotte et al, 2013):

- The Paleocene-Eocene thermal maximum (~55Ma). Massive carbon release caused acidification of the ocean and the temperature is thought to have been 4°C–7°C warmer than pre-Paleocene time.

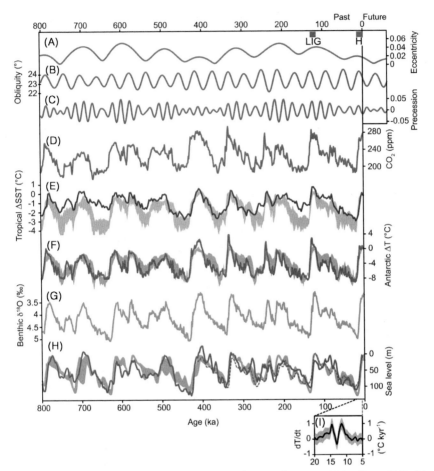

Figure 4.2 Orbital parameters and proxy records over the past 800 ka. Parts (BA), (b), and (C) show variations in eccentricity, obliquity and precession, respectively. Parts (D), (E), and (F) show atmospheric CO_2 concentration from Antarctic ice cores, tropical sea surface temperature, and Antarctic temperature based on ice cores, respectively. Part (G) shows benthic $\delta^{18}O$ (0), a proxy for ice volume, and part (H) shows reconstructed sea level. Shaded regions represent the range of simulations based on climate models. The last interglacial (LIG) is indicated at the top. The rate of change of global temperature during the last glacial termination is shown at the bottom. *(Source with permission: Masson-Delmotte et al., 2013.)*

- The Early Eocene climate optimum (50–52 Ma). There were no major polar ice sheets at this time and CO_2 concentration was approximately 1000 ppm. The temperature is estimated to have been 9°C–14°C warmer.
- The mid-Pliocene thermal maximum (3–3.3 Ma). Carbon dioxide levels are estimated to have been 350 to 450 ppm. Surface air temperatures were

1.9°C to 3.6°C warmer relative to the 1901–1920 average. Boreal forest moved north and replaced the tundra. This period may be analogous to our own climate by the end of the century (or earlier) but orbital forcing was probably different at that time compared to today.

4.4 Pleistocene glacial-interglacial dynamics

Fig. 4.2 shows the record of CO_2 back to 800 ka from Antarctica ice cores (Petit et al., 1999; Lüthi et al., 2008). Although orbital forcing (top of Fig. 4.2) is thought to be the main driver of glacial-interglacial cycles (e.g., Hays et al., 1976), the covariance of CO_2 with other proxies for climate (such as $\delta^{18}O$ in deep sea cores, sea level, and tropical sea surface temperature) indicates CO_2 plays an important role in glacial-interglacial dynamics. This occurs possibly through carbon cycle feedbacks involving the ocean. Methane also shows the same covariance with CO_2. Orbital forcing, however, appears to be the driving force for northern Hemisphere (NH) ice sheets due to changes in insolation. Fig. 4.2 shows an approximate 100 ka periodicity for glacial-interglacial cycles younger than about 430 ka suggesting the eccentricity component of the Earth's orbit is the causal factor. The precession and obliquity curves also show variations in intensity and periodicity and each glacial-interglacial cycles shows substantial differences to each other. The cause of the differences before and after 430 ka is not well understood but they are clearly visible in Fig. 4.2.

The sawtooth patterns in Fig. 4.2 indicate that the growth of ice sheets is slow whereas deglaciation was more rapid. As the ice expands over vegetation the albedo increases which is a positive feedback for ice sheet growth. Also, as the ice sheet grows and increases in elevation the atmosphere above contains less moisture producing less snow precipitation – a negative feedback. In the case of deglaciation, coupled thermo-mechanical processes appear to be at work on the ice sheets. High topography increases calving on the margins, and together with basal outflow, mass loss occurs. Massive ice melting leads to higher sea level and possibly further melting depending on the ocean temperature. The response time of large ice sheets is of the order of 10^4–10^5 years, so that they are not in equilibrium with current orbital forcing parameters. The implication of this for our present situation is that the full effects of warming on the cryosphere may not be seen for centuries or millennia.

Fig. 4.3 shows part of the methane record from the Greenland Ice Sheet Project (GISP2; Brook et al., 1996) during the Last Glacial Maximum (LGM,

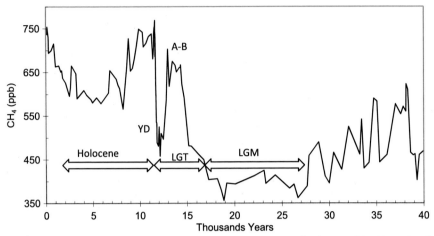

Figure 4.3 Methane concentration (ppb) over the past 40 ka from GISP2 (Greenland ice sheet project 2) ice core. A-B, the Allerød-Bølling interstadials; LGM, last glacial maximum; LGT, last glacial termination; YD, the Younger Dryas. *(Data source: NOAA.)*

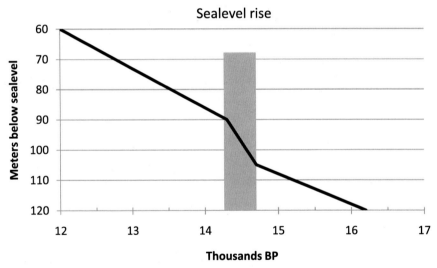

Figure 4.4 Sea-level rise during the Bølling warming at the end of the last glacial maximum. The shaded region is the melt water pulse 1A event. *(Source: redrawn after Deschamps et al., 2012.)*

28 ka–18 ka) and the Last Glacial Termination (LGT, 18 ka–11.5 ka). During the LGM the methane concentration was approximately 400 ppb (parts per billion) whereas during the LGT it was about 650 ppb (today it is about 1850 ppb, most of which is anthropogenic in origin). It is thought that the

increase in methane during the LGT was due to emissions from tropical wetlands as northern wetlands were likely still frozen (Brook et al., 1996).

During the LGM the global cooling is estimated to have been only negative 2.3°C but the North Atlantic SST is estimated to have been 10°C cooler compared to today. The cooling was larger over land than the oceans, a feature that has been modeled successfully by AOGCMs (Box 3.1). Dust in ice cores during the LGM indicates two to four times today's deposition, indicating a cold-dry climate. The radiative forcing of this dust is estimated to have been about negative $1Wm^{-2}$.

According to the IPCC-AR5 report (2013) during the LGT the global mean temperature increased by 3°C to 8°C (Masson-Delmotte et al., 2013). Ice melting occurred in two phases: one at 17.5 ka to 14 ka (see Box 4.1 below) and a second one at 13.0 ka to 10.0 ka. Temperature changes in the southern hemisphere presaged changes in the northern hemisphere and at least two causes are thought to be responsible for this: differences in inter-hemispheric ocean heat transport and the faster response of sea ice melting in the summer in the southern hemisphere. A long-standing question has been whether temperature increased before CO_2 and CH_4 concentrations increased or vice versa. As more high resolution studies of ice cores were undertaken, it now appears that there is little or no time lag between temperature and the GHGs. The GHG increases are attributed to climate carbon cycle feedbacks (Masson-Delmotte et al., 2013).

4.5 The CLIMAP Project

An important international attempt to characterize Pleistocene climate was the CLIMAP Project (acronym for Climate: Long-Range Investigation, Mapping and Prediction). Formed in 1971, it was funded by the National Science Foundation of the United States and other international organizations and headquartered at the Lamont-Doherty Geological Observatory. CLIMAP's first results were presented in a 1976 publication that showed the difference between surface temperature of the world's oceans for August at the last glacial maximum (18 ka). Sea surface temperatures were based on forams and other microfossil assemblages from deep sea cores (Fig. 4.5). This map shows several important features. In the North Atlantic region at 18 ka the Gulf Stream crosses the Atlantic further south and travels eastward to Iberia so that warm waters are not transported to high latitudes. Remarkably, James Croll (1821- 1890), a Scottish investigator, predicted such a pattern in his now famous paper (Croll, 1864). This pattern for the Gulf Stream

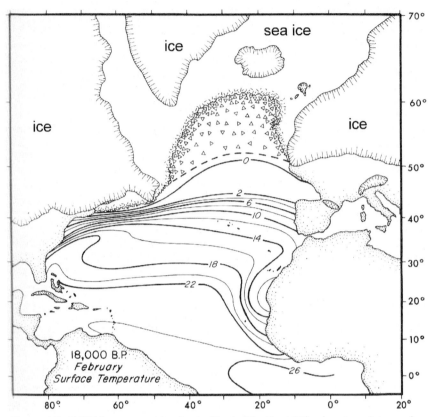

Figure 4.5 CLIMAP reconstruction for the North Atlantic at 18 ka showing winter surface temperatures (°C). Note the closely spaced isotherms for the east-west trending Gulf Stream at the latitude of Spain (40°N). Triangles indicate calved ice. *(Source: McIntyre, A., 1976. Courtesy Geological Society of America.)*

results in a temperature decrease of 8°C to 10°C (14.4 to 18°F) for the high latitude North Atlantic region compared to today. Large ice sheets are also indicated for North America and Northern Europe but surprisingly Alaska and large parts of Siberia were unglaciated. In the western Pacific the Sea of Japan also saw temperature decreases of 8°C to 10°C. Equatorial Pacific saw temperature decreases as much as 6°C. The global mean sea surface temperature was only 2.3°C lower than today, a surprisingly low value. Although controversial, the most recent value by the Intergovernmental Panel on Climate Change (IPCC-AR5, 2013; Table 5.2 of that publication) gives a value of 0.7°C to 2.7°C, which is consistent with the CLIMAP value.

The theory of orbital variations as the driver of Pleistocene ice ages was finally resolved by an interdisciplinary group of authors with expertise in microfossil paleontology, oxygen isotope stratigraphy, and advanced mathematical techniques. An important paper that resulted from this work was entitled "Variations in the Earth's Orbit: Pacemaker of the Ice Ages (Hays et al., 1976). Workers on the CLIMAP project were familiar with several hundred deep sea sediment cores. One of these workers at Lamont, J. D. Hays, recognized that in order to evaluate the astronomical hypothesis that a core must have several properties, namely: it should be continuous for about 450,000 years to be statistically useful; it must have accumulated at a fast enough sedimentation rate so that bioturbation had not disturbed the short period signals and it should be far from the influence of continental sediments. The only two cores to meet these criteria were from the southern Indian Ocean. Treating the core signals as a time series and applying mathematical techniques (spectral analysis and band pass filter analysis) to the geological and isotopic signals they identified the three Milankovitch periods corresponding to precession of the seasons (23,000 years), axial tilt (41,000 years) and eccentricity (100,000 years). A hundred years after Croll addressed this problem, it was finally solved but not without hundreds of deep sea sediment cores from all of the world's oceans were examined by hundreds of scientists worldwide. The CLIMAP Project therefore indirectly played an important role in this crucial advance.

4.6 Holocene climate

4.6.1 The 8.2 ka cold event

Compared to the Pleistocene epoch, the Holocene epoch (11.5 ka to the present) was warmer and wetter and more stable, with the exception of the 8.2 ka event. This event lasted only 150 years (compared to one thousand years for the earlier Younger Dryas, YD; Fig. 4.3) but it shows similarities to the YD suggesting a similar cause, namely disruption of the North Atlantic Gulf Stream due to melt-water from breakup of the northern continental ice sheets (Alley et al., 2015). The event is clearly visible on the methane record from the GISP2 Greenland ice core where methane drops from about 725 ppb to about 555 ppb (Fig. 4.3). Dry conditions, indicated by enhanced dust levels in Greenland ice cores, would have reduced wetland sources for methane production. Based on oxygen isotope analysis of the ice cores ($\delta^{18}O_{ice}$), the temperature is estimated to have dropped by negative $6 \pm 2°C$ (Alley et al., 2015).

There is substantial archaeological evidence that this cold and dry event had different impacts on late Neolithic societies in different regions. A well-dated archaeological site in northern Syria (Tell Sabi Abyad) spans the 8.2 ka events and shows these farmers adapted to the colder conditions – sheep and goats appear, which replace pigs that are not suited to arid conditions. A substantial increase in wool production is indicated by increased use of spindles, presumably for production of warmer clothing. New types of pottery were introduced possibly for food storage. All these changes occurred around the 8.2 ka event and ^{14}C dates show continuity across this event (van der Plicht et al., 2011). This event was not catastrophic for these people and they showed the ability to cope and adapt to the new conditions. In contrast, a different picture is seen further to the north on the west coast of Scotland where a hiatus in ^{14}C dates is interpreted as a retreat of the entire population from this area at this time (Wicks and Mithen, 2014). The climate conditions in the North Atlantic area were likely much more severe compared to the Middle East. The 8.2 ka cooling event appears to have been global in its extent and its brevity underscores the possibility of severe rapid climate change on a century time scale (Alley et al., 2015).

4.6.2 Last 2000 years

Fig. 4.6 shows an example of the global spatial distribution of different proxies for climate for three time periods: 1750 AD, 1500 AD, and 1000 AD. The proxies include tree rings, ice cores, bore holes, and other low resolution records and not surprisingly, the record becomes more sparse back in time. Some of these data are used as input in Atmosphere-Ocean General circulation models (AOGCMs) for the past 2000 years in the northern Hemisphere based on several different radiative forcings including solar, volcanic, GHGs, orbital, aerosols, and land use change. One conclusion from these simulations is that the last 30 or 50 years in the northern hemisphere were warmer than any 30 or 50 year mean during the last 800 years (Masson-Delmotte, 2013).

Two well known climate anomalies are the Medieval Climate Anomaly (950 AD to 1250 AD) and the Little Ice Age (1450 AD to 1850 AD). These periods of warm and cold climate, respectively, were recognized as early as the early twentieth century (IPCC-AR5, 2013; Table 5.1 in that publication). Large areas of Iceland were being cultivated in the 10th century and at the same time, Norse settlers colonized parts of Greenland. Voyages into ice-free areas further north, not previously navigable, were possible. More recent studies indicate the timing of this warm period was different in different areas and was not globally synchronous. The Little Ice Age saw expanded glaciers

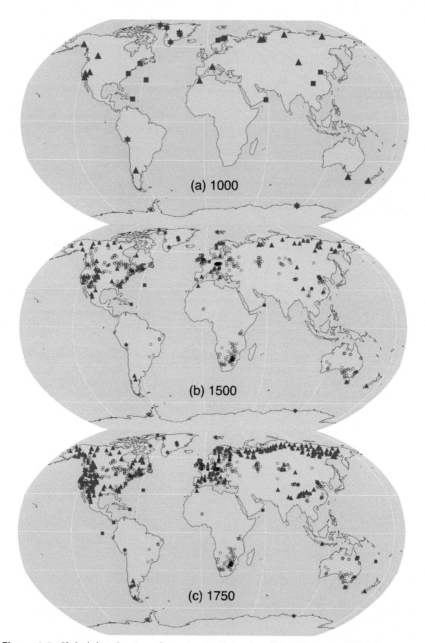

Figure 4.6 Global distribution of proxy records back to (A) 1000 years, (B) back to 1500 AD, and (C) back to 1750 AD. Triangles: tree rings. Circles: bore holes. Stars: ice cores (Greenland and Antarctica). Squares: other proxies. *(Source with permission: IPCC4 (2007).)*

in the Alps and Atlantic icebergs expanded southward and rivers froze in England (e.g., the Thames) and Europe.

References

Alley, R.B., Mayewski, P.A., Stuiver, M., Taylor, K.C., Clarck, P.U., 2015. Holocene climatic instability; a prominent, widespread event 8200 years ago. Geology 25, 483–486.

Bradley, R.S., 1999. Paleoclimatology: Reconstructing the Climates of the Quaternary. Elsevier, Boston.

Brook, E.J., Sowers, T., Orchardo, J., 1996. Rapid variations in atmospheric methane concentrations during the past 110,000 years. Science 273, 1087–1091.

Croll, J., 1864. On the physical cause of the change of climate during geological epochs. Phil. Mag. 28, 121–137.

Deschamps, P., Durand, N., Bard, E., Hamelin, B., Camoin, G., Thomas, A.L., Henderson, G.M., Okuno, J., Yokoyama, Y., 2012. Ice-sheet collapse and sea-level rise at the Bølling warming 14,600 years ago. Nature 483, 559–564.

Hays, J.D., Imbrie, J., Shackleton, N.J., 1976. Variations in the Earth's orbit: pacemaker of the Ice Age. Science 194, 1121–1132.

Hegerl, G.C., Crowley, T.J., Hyde, W.T., Frame, D.J., 2006. Climate sensitivity constrained by temperature reconstructions over the past seven centuries. Nature 440, 1029–1032.

Imbrie, J., Imbrie, K. P., 1986. Ice ages: solving the mystery. Harvard University Press.

IPCC4, 2007. Climate change: The Physical Science Basis. In: Solomon, S., Qin, D., Manning, M., Marquis, M., Averyt, K., Tignor, M.M.B., Miller, H.L., Chen, Z. (Eds.), Contributions of Working Group I to the fourth Assessment Report of the Intergovernmental Panel on Climate Change. Cambridge University Press, UK and New York.

IPCC5, 2013. Climate change: the physical science basis. In: Stocker, T.F., Qin, D., Plattner, G.K., Tignor, M., Allen, S.K., Boschung, J., Nauels, A., Xia, Y., Bex, V., Midgley, P.M. (Eds.), Contributions of Working Group I to the Fifth Assessment Report of the Intergovernmental Panel on Climate Change. Cambridge University Press, UK and New York.

Jensen, E., Overpeck, J., Briffa, K.R., Duplessy, J.-C., Joos, F., Masson-Delmotte, V., Olago, D., Ott-Bliesner, B., Peltier, W.R., et al., 2007. Paleoclimate. The physical science basis. In: Solomon, S., Qin, D., Manning, M., Marquis, M., Averyt, K., Tignor, M.M.B., Miller, H.L., Chen, Z. (Eds.), Contributions of Working Group I to the Fourth Assessment Report of the Intergovernmental Panel on Climate Change. Cambridge University Press, UK and New York.

Lüthi, D., Le Floch, M., Bereiter, B., Blunier, T., Barnola, J.-M., Siegenthaler, U., Raynaud, D., Jouzel, J., Fischer, H., Kawamura, K., Stocker, T., 2008. High-resolution carbon dioxide concentration record 650, 000–800, 000 years before present. Nature 453, 379–382.

Masson-Delmotte, V., Schulz, M., Abe-Ouchi, A., Beer, J., Ganopolski, A., González Rouco, J.F., Jansen, E., ten others, 2013. Information from Paleoclimate Archives. In: Stocker, T.F., Qin, D., Plattner, G.K., Tignor, M., Allen, S.K., Boschung, J., Nauels, A., Xia, Y., Bex, V., Midgley, P.M. (Eds.), The Physical Science Basis. Working Group I Contribution to the Fifth Assessment Report of the Intergovernmental Panel on Climate Change. Cambridge University Press, UK and New York.

McIntyre, A., 1976. Glacial North Atlantic 18,000 years ago: a CLIMAP reconstruction. Geol. Soc. Am. Mem. 145, 43–76.

Petit, J.R., Basile, I., Leruyuet, A., Raynaud, D., Lorius, C., Jouzel, J., Stievenard, M., et al., 1999. Four climate cycles in Vostok ice core. Nature 387, 359–360.

Van der Plicht, J., Akkermans, P., Nieuwenhuyse, O., Kaneda, A., Russell, A., 2011. Tell Sabi Abyad, Syria: radiocarbon chronology, cultural change and the 8.2 ka event. Radiocarbon, 53, 229–243.

Wicks, K., Mithen, S., 2014. The impact of the abrupt 8.2 ka event on Mesolithic population of western Scotland: a Bayesian chronological analysis using activity events as a population proxy. J. Arch., Sci. 45, 240–269.

PART II

5. Climate impacts: US sectors and regions   81
6. Adaptation   105
7. Mitigation   123

CHAPTER 5

Climate impacts: US sectors and regions

5.1 Introduction

This chapter is largely a summary of the Fourth National Climate Assessment, volume 2, titled Impacts, Risks and Adaptation in the US (USGCRP, 2018). The term impacts refers to the effects of climate events and climate change over a specific time frame on human and natural systems and includes the effects on lives, livelihoods, health, ecosystems, economies, societies, and cultures (IPCC-AR5, 2014, Working Group II). The impact of climate change on floods, droughts, and sea level rise is a subset of impacts called physical impacts. The first part of the chapter examines climate change impacts in key sectors such as water, energy, forests, coastal areas, land cover change, ocean and marine resources, agriculture, and cities and the urban environment. Under climate change, these sectors interact and depend on one another (e.g., water and agriculture) leading to complex behavior that is difficult to predict so that a multi-sector perspective is needed. The second part of the chapter examines these effects in geographically distinct regions within the US, including the northeast, southeast, southwest, and Alaska. These regions have different climates, economies, and diverse populations and geographies, and are responding to climate change in different ways.

5.2 Key sectors

5.2.1 Freshwater

A reliable supply of clean freshwater to individuals, communities, and ecosystems is critical to human and ecological health. There are significant changes in water quantity and quality across the country. Increasing temperatures and variable precipitation are intensifying droughts, increasing heavy downpours (extreme events) and reducing snowpack (USGCRP, 2018a). Reduced snow-to-rain ratios leads to less surface storage and increased runoff resulting in less aquifer recharge, particulary in the western United States. Increasing temperatures are leading to earlier melting of snow and less winter snow precipitation, altering hydroelectric production schedules.

Climate Change in the Anthropocene.
DOI: https://doi.org/10.1016/B978-0-12-820308-8.00001-5

Agriculture requires more water for irrigation and livestock as temperatures rise. Less groundwater recharge exacerbates droughts and provides less water for irrigation. Groundwater provides more than 40 percent for use by agriculture (livestock and irrigation) and domestic supply. The largest aquifer in the central United States, the High Plains aquifer, shows large depletions, especially in its southern portions (see Box 5.1).

BOX 5.1 The Ogallala aquifer.

The High Plains Ogallala aquifer is one of the largest in the world underlying 450,000 km^2 in eight Midwestern states (McGuire, 2017). There is substantial climate variation from north to south with a mean annual temperature in the north of 4°C and in the south 18°C. The western side also has a lower mean annual rainfall of 30 cm (25 cm per year represents desert conditions) and the eastern side, 84 cm per year. Because of high temperatures and winds, evaporation rates are high in the south (Texas and New Mexico) and this leads to deposition of a calcium carbonate crust (caliche), which inhibits aquifer recharge. The geologic units of the aquifer range in age up to 6 million years old and belong to the Tertiary and Quaternary geologic periods and consist of clay, silt, sand, and gravel. In the north, Nebraska is underlain by ancient sand dune deposits and this area is an important recharge area for the entire aquifer.

More than 95% of the groundwater goes to agriculture and the remainder for domestic use. The population of the region has increased from about 1 million in the 1900s to about 2.3 million today. Dust bowl conditions existed in the 1930s in Texas. Soil management strategies and water from the aquifer have since helped stabilize the soil. The economy of the region today depends almost entirely on irrigated agriculture. Twenty percent of the nation's wheat, corn and cotton come from the region and the southern and central parts account for one third of the nation's beef cattle production. In 2007, agricultural products from the region amounted to $35 billion.

The aquifer is now regarded as a nonrenewable resource as current extraction far exceeds recharge and excessive pumping has led to major declines in the water level (up to 46 m in places), especially in the central and southern portions (Fig. 5.3). Parts of the northern aquifer show some relatively minor rises in water level. Drought-tolerant crop varieties, precision irrigation technologies, and weather-based irrigation scheduling tools help conserve groundwater resources. Climate change, especially in the form of drought, will nevertheless represent a major challenge to the health of the aquifer.

Climate change is the primary driver of increased frequency and duration and strength of storms and other extreme events. Fig. 5.1A shows the number of extreme events (tropical cyclones, floods, droughts) over the period

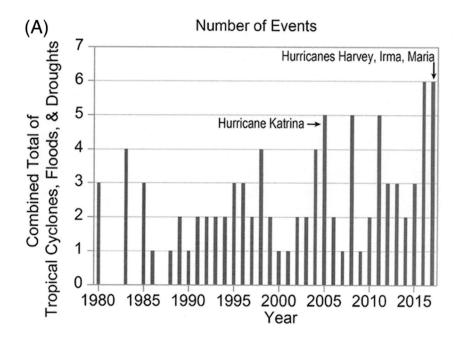

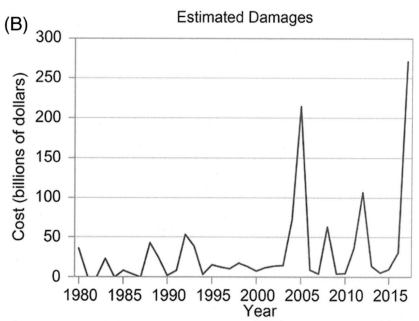

Figure 5.1 (A) The number of extreme weather events (tropical cyclones, floods, and droughts) over the period 1980–2015. Some major recent hurricanes in the southeast are indicated. (B) Cost of damages in billions of dollars over the same period. *(Source: NOAA, 2018.)*

1980–2015 and part (B) of this figure shows the cost in billions (10^9 $US) for damages with major peaks corresponding to recent hurricanes (e.g., Katrina, Harvey, Irma and Maria). As the population increases in coastal regions, these costs will only increase. More heavy downpours mobilize pollutants, sediment, and nutrients, resulting in lower water quality.

Deteriorating water infrastructure (including dams, levees, sewers, water distribution systems and waste water treatment plants) is more susceptible to failure during extreme events but also fail during normal events. Much of this infrastructure in the US is past or close to its planned life expectancy. The replacement and repair cost of this infrastructure will be a substantial fraction of the entire GDP (about $20 trillion in 2020). Water management strategies by private, federal, and public agencies (e.g., water utilities or the U.S. Army Corps of Engineers) need to take into account future climate scenarios that are outside those seen in the twentieth century. These plans should include a number of factors such as updated engineering design, water quality regulations, insurance programs, and water allocation to name a few. President Biden proposed a two trillion (10^{12} $US) infrastructure plan in 2021.

5.2.2 Energy supply, delivery, and demand

A powerful storm from the northeast with winds of up to 90 miles an hour hit New England in October 2019. The pressure drop (1000 millibars to 975 millibars) associated with the cyclone was the largest seen since records began in 1872. Half a million people in New England lost their power, mainly due to tree falls. A few days earlier, the major electrical utility in California deliberately cut power in areas that were subject to drought in order to avoid igniting wildfires in the forest-brush caused by sparking power lines. In the Gulf of Mexico, drilling and production oil rigs are commonly shut down when major storms approach. Hurricane Irma cut electricity to two thirds of the population of Florida in 2017. These examples show that extreme weather events associated with climate change can have a major impact on the production and delivery of energy (USGCRP, 2018b). The resilience of the energy infrastructure must be able to withstand likely future climate scenarios since it underpins almost every sector of the economy, including manufacturing, agriculture, transportation, healthcare, banking, and telecommunications.

There are many other examples of how climate affects the energy sector. Rising temperatures will increase demand for air conditioning but are only partly offset by lower heating demands during warmer winters. Flooding of roadways and freight railways due to extreme precipitation events can shut down delivery of fuel, food, and other goods. Power plants on the coast are vulnerable to flooding from rising sea level and hurricanes. During drought, lower water levels reduce the capacity of power plants (thermoelectric, nuclear, and hydroelectric) all of which need large amounts of cooling water.

In the United States, electricity generation by source for 2018 was as follows (in percent): natural gas, 35; coal, 27; nuclear, 19; hydropower, 7; wind, 7 and solar, 2. The most important recent changes in these sources was a decrease in the use of coal and an increase in use of natural gas; in addition, renewable sources have become more cost competitive. New energy technologies can help increase the resilience of the energy system at the local, state, and federal levels in the face of climate change. Needless to say, an additional goal should be to reduce emissions especially in the transportation and electricity generation sectors.

5.2.3 Land cover and land-use change

Changes in land cover influence local to global scale weather and climate patterns by altering the flow of energy (e.g., change in albedo) and water budget (through evaporation and precipitation) and greenhouse gas transfers (CO_2 and CH_4) between the land and the atmosphere (USGCRP, 2018c). Reforestation can produce local cooling and conversely deforestation can reduce rainfall, promoting drought conditions. Reduced precipitation can convert forest to grassland and this may increase albedo. On the other hand, urban areas tend to have lower albedo producing the urban heat island effect. Changes in land-use and land cover occur in response to both human and climate drivers. Land-use and land cover are inherently coupled – land cover change enables specific land-uses; conversely changes in land-use practices can result in land cover change. For example, manmade fires in the Amazon rainforest clear the forest for agricultural use. Such land-use decisions are typically based on short-term economic factors. The recent expansion of oil and gas extraction over large areas in the US (i.e., hydrofracturing) is an example of land-use change and cover change based on several factors: policy, economics and technology. Remote sensing by satellite can provide information on land use and land cover on a global scale, but at a relatively low resolution (e.g., km^2 scale).

5.2.4 Forests

Forests occupy 33% of land area in the US. It is very likely that climate change will decrease the ability of many forest ecosystems to provide important ecosystem services to society. These services include timber

Table 5.1 Employment, GDP, and population for shore adjacent counties.

Region	Employment Millions %	GDP Trillions %	Population Millions %
US	134	16.7	327
Shore counties	50.2% 37.5%	7.2% 43.2%	118% 37.4%

Source: USGCRP, 2018e. Data for 2016.

supply (private forests), carbon sequestration, water quality and quantity, recreation, and fish and wildlife habitat (USGCRP, 2018d). Increasing temperature, more frequent droughts and forest disturbances (wildfires, insect outbreaks) are expected to decrease tree growth and carbon storage in most locations. Forests store very large amounts of carbon and reforestation is therefore encouraged in order to mitigate carbon emissions; conversely, deforestation (biomass burning) results in release of stored carbon. Wildfires burned 3.7 million acres nationwide from 2000 to 2016 and tree mortality due to drought caused the loss of 300 million trees in Texas in 2011, and 129 million trees in California between 2010 and 2017. Between 2010 and 2015 the federal government spent between $0.8 billion and $2.1 billion on fire suppression.

5.2.5 Coastal areas

United States coasts span three oceans plus the Gulf of Mexico, the Great Lakes and Pacific, and Caribbean islands. America's coastal property and public infrastructure is worth $1 trillion and is vulnerable to the increased frequency, depth, and extent of tidal flooding due to sea level rise and storm surges (USGCRP, 2018e). The 60,000 miles of floodplain roads and bridges are already vulnerable to hurricanes and storms costing billions of dollars in repairs, not including indirect costs such as lost business. The cost of such events has a cascading effect on the larger economy. Even under the lower representative climate pathways (RCP2.6 and RCP4.5; Chapter 3) flooding costs will increase and property values will decrease. Under the higher pathway (RCP8.5, somewhat similar to "business as usual") coastal communities will be transformed by the latter half of the century.

Coastal counties made up 42% of the US population at 133 million in 2013. Employment is in a wide variety of areas including fishing, defense, transportation, and tourism and the coasts are active hubs of commerce that connect to our trading partners worldwide. Table 5.1 presents some data for shore adjacent counties in terms of employment, population, and GDP as a percentage of the entire US coastal regions are made up of a diverse set of ecosystems, including beaches, intertidal zones, reefs, sea grasses, salt marches, estuaries, barrier islands, and deltas that support a range of ecosystem services, including fishing, tourism, recreation, and coastal storm protection.

Coastal property owners are likely to bear the costs from sea level rise and storm surge, including storm damages, property abandonment, protection measures such as property elevation, beach nourishment, and coastline armoring. Residents in flood hazard zones are required to buy flood insurance, which is usually federal insurance as few private policies are available. It is not uncommon for repeatedly flooded properties to be rebuilt several times, at great cost to the taxpayer; relocation should be encouraged.

5.2.6 Oceans and marine resources

Ocean ecosystems around the US provide a wide variety of goods and services, such as food, jobs (1.6 million), recreation, and energy. The fishing sector alone is worth $200 billion of economic activity. Ocean ecosystems include coral and oyster reefs, kelp forests, mangroves, salt marshes, and provide habitat for many species as well as protection of the shoreline from storms (USGCRP, 2018f). In addition, many of these ecosystems have the capacity to sequester carbon.

The oceans play a critical role in global climate systems as they redistribute heat (mainly from the tropics toward the poles) and they absorb carbon dioxide from the atmosphere. It is estimated that the oceans have absorbed 30 percent of anthropogenic CO_2 emissions since industrial times (Rhein et al., 2013). It is also estimated that the oceans have absorbed 90 percent of the heating associated with these CO_2 emissions; otherwise the average global air temperature would be higher. Carbon dioxide atmospheric emissions have three major influences on the oceans: they cause ocean warming and acidification, and also deoxygenation. Between 1900 and 2016 the global mean surface temperature of the oceans has increased by $0.7°C \pm 0.8°C$. The warming oceans have several impacts including: sea level rise (due to thermal expansion of seawater), stratification (vertical density gradients), which in turn affect ocean circulation patterns and also ocean productivity (e.g., plankton growth) and species migrations (typically northward and into deeper waters). These latter migrations have important implications for fisheries including higher costs for different fishing gear, fishing in deeper water, and longer trips to fishing grounds.

Since the beginning of the industrial revolution the pH of the surface ocean water has decreased by about 0.1 units from 8.2 to 8.1 (corresponding to an increase in acidity). Since pH is a logarithmic scale this corresponds to an increase in hydrogen ion concentration of 25%. As carbon dioxide from the atmosphere dissolves in the oceans, carbonic acid (H_2CO_3) is produced which in turn beaks down into bicarbonate (HCO_3^-) and H^+ ions, thereby increasing the acidity:

$$CO_2 + H_2O \leftrightarrow H_2CO_3 \leftrightarrow HCO_3^- + H^+ \tag{5.1}$$

An additional reaction also consumes carbonate ions (CO_3^-) to produce more bicarbonate:

$$CO_3^- + H^+ \leftrightarrow HCO_3 \tag{5.2}$$

The two-headed arrows indicate these reactions are reversible. The addition of carbon dioxide to the oceans therefore increases acidity (decreases pH) and consumes carbonate ions. If the concentration of carbonate ions drop below the saturation level, dissolution of $CaCO_3$ (either as calcite or aragonite) occurs reducing the ability of marine organisms to build their skeletons and possibly their viability. Decreasing thickness of exoskeletons in marine carbonate secreting organisms has already been observed.

Deoxygenation of the ocean has occurred due to the decrease in the solubility of oxygen with higher ocean temperatures. Warmer surface water also increases stratification so that less oxygen is transported to deeper levels. Oxygen in all the tropical ocean basins has decreased over the past 50 years. Between 1970 and 1990 the mean annual global oxygen loss between 100 m and 1000 m depth is calculated to be $0.55 \pm 0.13 \times 10^{14}$ moles per year (Rhein et al., 2013). Additional deoxygenation can occur due to agricultural runoff of nitrogen and phosphorous. These fertilizers cause plankton blooms which when they die are consumed by bacteria using dissolved oxygen. This leads to hypoxic zones (or dead zones) when oxygen falls to less than 2 ppm, threatening marine life. One of the largest of these zones is offshore Texas and Louisiana in the northern Gulf of Mexico with an area of 20,000 km^2. Flooding in upstream basins has the effect of flushing out more nutrients, increasing the size of the dead zone. Lake Erie, one of the Great Lakes, also has a large dead zone, with the same agricultural cause.

A relatively new phenomenon is that of marine heat waves (also called heat blobs), which are likely due to climate change, where large areas of the ocean have temperatures up to 2°C warmer than normal (not predicted to occur for the oceans by models until the end of the century). One such

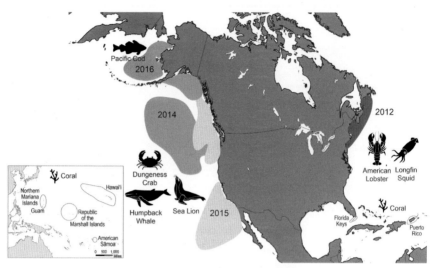

Figure 5.2 Extreme weather events in US waters since 2012. The heat wave of 2012 in the Gulf of Maine is indicated that affected American lobster and longfin squid. Several heat waves occurred in the North Pacific and the Gulf of Alaska over the period 2014–2016 and affected Pacific cod, crab fisheries and indirectly led to the death of sea lions and whales. Coral reefs that experienced moderate to severe bleaching are indicated by the coral icon. *(Source:USGCRP, 2018f.)*

blob occurred off the coast of Maine in 2012 and much of the west coast was affected during 2014–2016 (Fig. 5.2). More recently (2019) a blob appeared offshore of the Hawaiian islands, causing bleaching of coral reefs which will likely reduce fish habitats. In the case of the Gulf of Maine blob, the lobsters there spawned earlier than usual, thereby disrupting the lobster industry, causing prices to collapse.

The physiology and performance of marine species sensitive to warming, acidification, and deoxygenation can be directly affected. Changes in their relative abundance (decline or increase) in an ecosystem may occur if they relocate or colonize new locations. Such reorganization of species can result in their losing resources such as prey and shelter. Other species may be exposed to new conditions they have not encountered before. Reorganization of such communities would change the ecosystem services and influence regional economies, fisheries harvest, and aquaculture.

These impacts are observed worldwide but the most sensitive ecosystems are in tropical and polar regions. In the US, mass coral beaching has been observed offshore Puerto Rico, the Virgin islands, Florida, Hawaii, and US Pacific islands (see also Fig. 5.2). The extent of sea ice in the Arctic

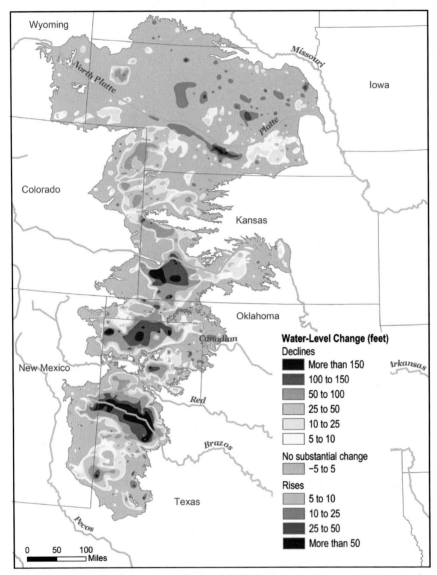

Figure 5.3 Water level changes in the Ogallala high plains aquifer up to 2015. Note the depletions in the southern portions of the aquifer. *(Source: McGuire, 2017.)*

is decreasing causing further ocean warming (the sea ice/albedo effect) leading to loss of habitat for polar bears and seals who use the ice for hunting, migration, and shelter. Under the high scenario (RCP8.5), most marine ecosystems (85%) by the year 2050 will experience a combination of temperature and pH conditions not seen by modern species. Warm water and acidification pose high risks for pteropods (sea snails), bivalves, krill, sea grasses, and corals. The ability of these organisms to adapt, and how fast, is not well understood. Fisheries management can help reduce impacts and promote resilience in the face of changing conditions.

5.2.7 Agriculture and rural communities

The changing climate presents real threats to US agricultural production, forest resources, and rural economies. Some of these threats are (USGCRP, 2018g):

- More severe storms: the year 2012 saw the most intense storms in recorded history.
- Rising average temperatures: encourages invasive species and increases the cost of weed and pest control. Livestock are also less productive with increasing temperature, as are many crops.
- Extreme precipitation events. The northeast region is seeing heavier rains while the southwest is seeing increasing drought, affecting nut, fruit, and vegetable production.
- More forest fires. The fire season in the US is 60 days longer than 30 years ago. It is estimated that the acreage burned will double by 2050.

United States farmers are among the most productive in the world. US farms contributed $137 billion to the economy (almost 1% of GDP) in 2015 accounting for 2.6 million jobs. A major proportion of rural communities depend on agriculture and related industries. Forty six million people (15% of the US population) live in rural areas, corresponding to 72% of the nation's land area. Feedstocks, livestock products, and horticultural products help support rural economies across the nation. In general, rural areas are showing slower employment growth and slower population growth rates, with higher poverty rates, compared to urban areas.

US agricultural production relies heavily on the nation's land, water, and other natural resources. Land use can change depending on climate change and agricultural practices. In 2012, 40% of the land area was farmland, with pasture accounting for 40%, cropland 43%, and woodland 8%. Bioenergy crop production is increasing, mainly due to ethanol production from corn, which made up about 10% of US motor fuel needs in 2019.

Higher temperatures play a major role in increasing drought intensity and drought onset — altering water availability and demand. Increased evaporation leads to increased plant stress, reduction in yield, and increased wildfire risk and also depletion of surface and groundwater resources. Soil carbon is depleted during drought due to lower plant productivity. In 2012, drought in 66% of US counties were declared disaster areas and the federal crop insurance program paid out $14.5 billion. Livestock, wheat, corn, and soybean production in the Great Plains and the Midwest were adversely affected by drought. A prolonged drought (2013–2016) in all of California depleted reservoirs and groundwater supplies with large declines in aquifer levels. In 2014, the California legislature passed the Sustainable Groundwater Management Act to plan management of these resources for the next 10–20 years.

Average yields of crops (e.g., corn, soybean, wheat, rice, oats, and silage) decline above certain maximum temperature thresholds especially during critical reproductive periods. Additionally, loss of synchrony between crops and pollinators presents a further challenge. Both increasing temperature and CO_2 concentration increase the competitiveness of weeds. Continued expansion of irrigation will be limited by water resources and the decreasing profitability of irrigated production. Wildfires will increase in several ecosystems (e.g., grasslands, rangelands, and forests) as the wildfire season becomes longer.

Degradation of soil and water resources is predicted to be an important outcome of climate change. Soil erosion is an important environmental threat to sustainable crop production. In addition, it also affects drainage networks, water quality, and recreation. The predicted increase in extreme precipitation events will result in increased soil erosion and increased sedimentation into water bodies, resulting in the loss of carbon and nutrients and a reduced agricultural productivity. Increased runoff will result in excess nutrients being transported to drainage basins producing hypoxia zones in adjacent water bodies. There is a predicted increase in the frequency, duration and intensity of these dead zones (see Oceans and marine resources above). Soil protective measures include grassed waterways, cover crops and conservation tillage. Increasing temperatures are predicted to threaten the health of humans working in agriculture and that of livestock.

5.2.8 Cities and urban environment

Urban areas, where in 2015, 85% (275 million) of the US population live and where 90% of the GDP is produced. The five largest cities alone produce 23% of GDP. By 2100 it is projected between 425 and 696 million people will live in metropolitan areas. Urbanization is therefore increasing which affects air, water, and soil quality. Increases in impervious surface cover (cement and asphalt) increase the likelihood of flooding especially in coastal cities. In many cities, infrastructure (such as residential and commercial buildings, transport, communications, energy, water systems, and parks and streets) is nearing the end of its planned service life and repairing deteriorating infrastructure is estimated to cost $3.9 trillion by 2025 (USGCRP, 2018h). Older infrastructure is more likely to fail during extreme climate events and current infrastructure and building design do not take into account future trends in climate change. Climate susceptibility for urban populations varies by neighborhood, type of housing, age, occupation, and daily activities. People exposed to weather- and climate-related disasters are subject to adverse psychological effects, including depression, anxiety, and post-traumatic stress disorder, with the most vulnerable members of the population being children, the elderly, the economically disadvantaged, and the homeless.

Cities are places where people live together to work, learn, socialize, and recreate. Climate change impacts exacerbate existing challenges to urban quality of life and adversely affect urban health and well-being. Challenges to prosperity in city life include social inequality, deteriorating infrastructure, and stressed ecosystems. Populations experiencing social inequality or poor health have greater exposure and susceptibility to climate change. Renters and the homeless, for example, may have insufficient access to air conditioning and insulation, and have greater exposure to heat stress. Similarly, the elderly, children playing outdoors, construction workers, and others working outside are vulnerable to extreme heat. Air and water quality issues also affect the health of urban populations including the ease with which diseases may spread among high-density populations. Economic impacts of climate change include food price volatility and costs of energy, water, and insurance.

Coastal city flooding can result in forced evacuation, as seen recently in Houston, Texas (2017), and Charleston, South Carolina (2019), affecting family and neighborhood stability in addition to mental and physical health. Inland communities that take in evacuees are also affected. As seen in New Orleans after hurricane Katrina in 2005, thousands of evacuees have still not returned.

Current investments in the US are not sufficient to cover needed repair and replacement of critical infrastructure (USGCRP, 2018h). Infrastructure needs to perform reliably throughout the length of its life span and current infrastructure was not designed for projected climate extremes and is therefore more vulnerable to these changes. Above and below ground transportation systems are vulnerable to flooding, as seen in Manhattan after super-storm Sandy in 2012. Higher temperatures increase the stress on cooling systems and upgrades on buildings and the electric grid are needed to handle these temperatures. Sea level rise together with more frequent and intense storms may increase flooding in coastal cities rendering some buildings and public infrastructure unusable (see Coastal impacts above). Cities are centers of production and consumption of services and goods and they are enmeshed in regional and global supply chains. They also rely on local services for telecommunications, energy, water, healthcare and transportation to name a few. Gradual and abrupt climate change can disrupt these systems.

5.2.9 Human health

Extreme climate events such as drought, wildfires, heavy rainfall, floods, storms, and storm surges can all adversely affect human mental and physical health (USGCRP, 2018i). The effects can be exacerbated by pre-existing medical conditions, giving rise to adverse mental health effects such as anxiety, depression, post-traumatic stress, and suicide. These mental health impacts may combine with other health, social, and environmental stressors. Extreme weather events can also disrupt healthcare systems themselves. The 2017 hurricane season demonstrated the vulnerabilities of populations in Puerto Rico, the US Virgin islands, and other Caribbean islands.

Temperature extremes affect different groups in the population: with children, older adults, and pregnant women being more susceptible. Higher temperatures occur in urban areas compared to rural areas due to the urban heat island effect (concrete structures and less open space). Heat-related deaths outweigh those due to cold-related deaths for most regions. Analyses of hospital admissions and emergency room visits indicate heat-related health effects include kidney failure and cardiovascular and respiratory complications, electrolyte imbalance and preterm births.

Climate change will likely alter the geographic range, seasonal distribution, and abundance of vector-borne diseases, exposing North Americans to ticks and mosquitoes that carry a variety of viruses. One effect has been for virus-carrying mosquitoes from South America to move north into Florida and Texas. El Niño warm events serve to indicate the extent to which these changes will likely occur with further warming. Economic development may reduce some populations' exposure to these disease vectors. Increased water temperature is expected to alter seasonality of growth and range of harmful algae and pathogens (viruses, bacteria, fungi and parasites). More intense and frequent rainfall will increase runoff introducing pathogens and toxic algal blooms to water bodies. Cities with combined sewage and storm systems commonly overflow into nearby water bodies during heavy rainfall, introducing water-borne pathogens. This may lead to gastrointestinal illness for those without access to potable water. Increasing temperatures and extreme temperature events will expose food to pathogens and toxins (e.g., salmonella and other bacteria and fungi). The impact on human health will depend on food chain management, human behavior, and regulatory governance. Increase in wildfire frequency and size and a longer wildfire season degrades the air quality with smoke, increasing health risks, such as respiratory illness.

5.3 Regional climate impacts

Risks posed by climate change vary by key sectors as outlined above. Several sectors may be impacted simultaneously multiplying the effects that are difficult to predict. Climate change effects also vary substantially by region and this section summarizes four regions within the US social, economic, and geographic factors shape the exposure of people and communities to climate-related impacts. As mentioned already, risks are often highest for those already vulnerable: low-income communities, children, and the elderly. The four regions briefly addressed below are the northeast, southeast, southwest, and Alaska.

5.3.1 Northeast

This region encompasses states from Maine to Virginia (USGCRP, 2018j). The distinct seasonality of the northeast with cold winters and warm to hot summers and a diverse natural landscape provides the economic and cultural foundation for many rural communities in this region. These communities are supported by a diverse range of agriculture, tourism, and natural resource-dependent industries (forestry, fishing, aquaculture). Less distinct seasons with milder winters and earlier spring are already altering ecosystems and environments that are adversely affecting tourism (e.g., reduced snowpack), farming, and forestry. This region already has seen increases in intensity in rainfall above all other regions in the US leading to flooding and increased soil erosion.

The region's extensive coastal areas support commerce, tourism, recreation, and fishing which are important to the region's economy and way of life. The coast itself is highly varied and ranges from hard rock bluffs, to barrier islands, estuaries, and developed areas that will respond to rising sea level and storm surges in different ways with greater damage to fragile coastal ecosystems. Warmer ocean temperatures, sea level rise, and ocean acidification threaten many of the areas industries (see Ocean and marine resources above). Many fish and invertebrate species have moved northward and to greater depths on the continental shelf, adversely affecting fisheries. Much of the infrastructure in the northeast (power supply, transportation, water drainage, and sewer systems) is nearing the end of its planned lifespan and will be more susceptible to climate-related disruptions.

5.3.2 Southeast

This region includes Kentucky in the north and Florida in the south and the several states in between (USGCRP, 2018k). Geographically, it includes an extensive low-lying coastal region and a very large low-lying inland region, including several major metropolitan areas, and it also includes higher elevation terrain in the southern portion of the Appalachian mountains. It includes several coastal cities (from south to north), such as Miami, Jacksonville, Charleston, Savannah, and Norfolk. The coastal cities are vulnerable to multiple climate change risks, including flooding, storm surges, and infrastructure damage (see Coastal effects above). The population in the metropolitan areas is growing (e.g., Atlanta) while rural areas are losing population.

Cities in the southeast are experiencing longer and more intense summer heat waves, made worse by the urban heat island effect. Only five major cities in the US have increasing trends above the average for all aspects of temperature (e.g., frequency, duration), three of which are in the southeast (Birmingham, Rayleigh, and New Orleans). The number of days with nighttime temperatures above 75°C is increasing and the freeze-free season length has increased by 10 days over the 1900–2016 average. The number of extreme rainfall events has increased over the period 1900–2016 and the number of days per year with greater than 7.6 cm (3 in.) of precipitation is shown in Fig. 5.4. Under RCP4.5 and RCP8.5 scenarios these events are projected to increase. Flood events in Charlestown, South Carolina, have been increasing and by 2045 it is projected this town will have tidal flood events 180 times a year, compared to 14 for 2014. Four major flooding events

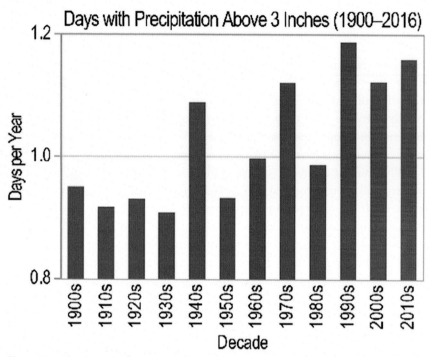

Figure 5.4 Number of days per year the southeast has experienced precipitation above 3 inches (7.6 cm) per year by decade. The frequency of such events has increased since the 1980s. *(Source: NOAA-NCEI.)*

have occurred in the southeast in the period 2014–2016, costing a total of $24 billion and over a hundred casualties (Fig. 5.1).

Infrastructure, such as roads, bridges, coastal properties, freight and passenger rail, water transport, and treatment facilities are all vulnerable to increased precipitation and extreme weather events. Several coastal cities have laid out plans for future sea level rise. Average global sea level has risen 20–23 cm (8–9 in.) since 1880, with 7.6 cm (3 in.) occurring since 1990. Tidal flooding is projected to increase causing road closures and infrastructure damage. This will impact tourism, transportation, and the economy as a whole. The potential cost of flooding in the southeast by 2050 is estimated at $60 billion under RCP8.5 and $56 billion under RCP4.5 scenarios.

Climate change is likely to modify the transmission, seasonality, and prevalence of vector-borne disease (e.g., from mosquitoes or ticks) in the southeast. These diseases pose a greater risk in cities than rural areas because of the denser population and more man-made water bearing structures. The southeast already has suitable conditions for virus-carrying mosquitoes from July to September. In Florida, these conditions already exist all year round.

Three major hurricanes made landfall in the US in 2017, mostly affecting the southeast and costing a total of $306 billion (Fig. 5.1). Hurricane Irma, a category 4 storm, tracked up the west coast of Florida with 1–1.5 m (3–5 ft.) of inundation which also affected large areas of Georgia and South Carolina. In Florida, 6.8 million people were ordered evacuated and 6.7 million people were without electricity after the storm passed. Most rivers in northern Florida, Georgia, and South Carolina were flooded.

The southeast's diverse natural systems which provide many benefits to society will be transformed by climate change. Changing winter temperatures, wildfires, sea level rise, hurricanes, floods and drought, and warming ocean temperature are expected to redistribute species both onland and offshore and modify ecosystems. Future generations can expect to interact with natural systems that will not be recognizable from those of today.

5.3.3 Southwest

This region includes Arizona, California, Colorado, New Mexico, Nevada, and Utah and it occupies 20% of the US land area (USGCRP, 2018l). Of a population of 60 million, nine out of 10 live in urban areas. The region extends across unique ecosystems from the Sonoran desert in the south to the Sierra Nevada mountains in the north. Approximately, one quarter of the region is underlain by forest and half is underlain by shrub-land. It provides the nation with half of its vegetables, fruits, and nuts and therefore is important to the food security of the US renewable energy resources, such as hydroelectric, wind, and solar account for 20% of the region's energy and it has a lower emissions output than the US average because of this. The US government oversees one half of this region as national parks and public lands, the original intention of which was to conserve wild life and plant life, clean waters, timber resources, and also to be used for recreation.

The California coast extends 5500 km (3400 mi.) and hosts several international airports, tens of thousands of homes and important seaports in Long Beach and Oakland. This coast is subject to multiple climate change vulnerabilities, including sea level rise, storm surges, ocean warming, and acidification, all of which are altering coastal ecosystems. Marine heat waves have caused the failure of some fisheries (Fig. 5.2). Because of the arid conditions, water resources (both surface and groundwater) are scarce with complex laws allocating water between agriculture, cities, fisheries, energy production, and between Mexico and the US.

Winter snowpack in the Sierra Nevada and the Rocky mountains and other mountains provide a source to much of the region's water. Three quarters of the region's water goes to irrigation. Spring snowmelt provides water to the Colorado, Sacramento, and Rio Grande and other major rivers, which is trapped by dams and transported long distances to major cities by aqueduct. Higher temperatures due to manmade climate change increase evaporation rates and reduce soil moisture and surface water supply. Higher temperatures have also reduced the snowpack with more precipitation falling as rain, and snowmelt is occurring earlier in the Spring. Groundwater levels are reduced due to rapid runoff (as rain rather than slower snowmelt) and excessive pumping for irrigation. Daily maximum temperatures regularly exceed 35°C (98°F) across the region. Average annual temperatures increased 0.9°C (1.6°F) between 1901 and 2016. The highest temperatures since 1895 occurred in 2014, 2015, 2016, and 2017. Populations vulnerable to hot temperatures as mentioned already are the young, the elderly and those with fewer economic resources, including Native Americans. Historically, indigenous people were relegated to reservations in drier areas; the southeast region is home to 186 Native American tribes.

Higher temperatures have contributed to recent severe droughts in the Colorado River basin, the Rio Grande and in California (Fig. 5.5). Increased aridification is due to higher temperatures, less soil moisture, less snowpack, and earlier snowmelting and evapotranspiration. Climate change in this region will result in a series of cascading and interacting effects that are difficult to predict.

5.3.4 Alaska

Alaska is the largest state in the nation – almost one fifth the size of the combined lower 48 states. The state has an abundance of natural resources, including oil, mining, and fishing and it also has a substantial tourist industry (USGCRP, 2018m). As part of the Arctic, Alaska is warming at twice the rate of the planet as a whole and is warming faster than any other state in the US average statewide temperatures for 2018–2019 were the second highest since 1900 when records began (Arctic Scorecard, 2019). Prior to the 1970s the average statewide temperatures were variable and did not show a significant trend. Since the 1970s, however, the average temperatures show an increase of 0.7°F (0.39 °C) per decade (Taylor et al., 2017).

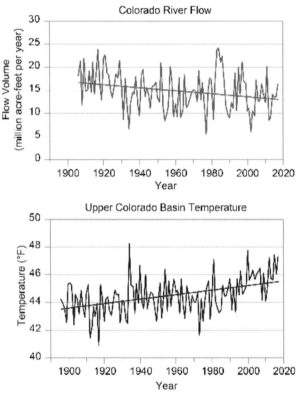

Figure 5.5 Upper panel. Decrease in Colorado River flow over the period 1910–2018. The decrease in flow has reduced lakes Powell and Mead to their lowest levels since initial filling. Lower panel. Increase in Upper Colorado River Basin temperature causing increased evaporation. *(Source: USGCRP, 2018a.)*

Due to warming, Arctic sea ice is now younger and thinner and the retreat of the ice has several implications for marine ecosystems and for indigenous people's way of life. Fish species are moving north with implications for commercial fisheries and subsistence fishing. Polar bears and walruses depend on sea ice as platforms for foraging, resting, and reproduction. Rural Alaska is inhabited by approximately 70 indigenous communities. Travel for these communities over thin ice is more dangerous and the ice precludes travel by boat further isolating these communities. The loss of coastal ice removes a platform for subsistence hunting and fishing and butchering of whales; on the other hand, new opportunities for hunting and fishing may open up under ice-free conditions and these communities have long shown resilience and adaptation abilities. Further warming of the oceans is occurring because of the sea-ice albedo effect and it is estimated the Arctic will be ice-free in the summer by mid-century. Retreat of coastal sea ice also increases coastal erosion allowing waves to reach the shoreline and increasing coastal flooding, requiring entire communities to relocate at great cost (GAO, 2009).

Increased atmosphere-ocean interaction due to melting ice increases ocean acidification (i.e., lower pH) with negative impact on calcium carbonate secreting organisms such as mollusks, crabs, corals, crustaceans and sea snails, species that are important to the food web in Alaskan waters. Parts of the seas around Alaska are already undersaturated with respect to aragonite (a polymorph of $CaCO_3$ common in shelled organisms) and other areas of the Arctic Ocean are expected to cross this threshold by 2030; under these conditions laboratory experiments indicate some organisms produce thinner shells and are less viable.

Thawing of the permafrost in Alaska due to warming has damaging impacts on infrastructure, causing roads to founder and buildings to collapse and estimates for the repairs to infrastructure range from \$110 to \$270 million per year. Warmer temperatures also contribute to more frequent and extensive wildfires that are already occurring. Permafrost in the Arctic region represents a substantial store of global carbon and thawing of permafrost releases CO_2 and CH_4 to the atmosphere furthering contributing to warming in a positive feedback cycle. The rapid warming of Alaska, and the Arctic in general, is likely partially due to the two positive feedbacks already mentioned – the seaice/albedo effect and thawing of the permafrost.

References

Arctic Scorecard, 2019. Arctic Council Conservation Scorecard. https://arcticwwf.org. Accessed July 22 2020.

GAO, 2009. United States Government Accountability Office. Alaska Native Villages. https://gao.gov/reports-testimonies. Accessed July 27 2020.

IPCC5, 2014. In: Field, C.B., Barros, V.R., Dokken, D.J., Mach, K.J., Mastrandrea, M.D., Bilir, T.E., Chatterjee, K.L., Ebi, K.L., Estrada, Y.O., Genova, R.C., Girma, B., Kissel, E.S., Levy, A.N., MacCracken, S., Mastrandrea, P.R., White, L.L. (Eds.), Summary for Policy Makers in Climate Change 2014, Impacts, Adaptation, and Vulnerability, Part A: Global and Sectoral Aspects. Cambridge University Press, Cambridge, UK.

McGuire, V.L, 2017. Water level and recoverable water in storage changes. High Plains Aquifer, predevelopment to 2015 and 2013 –15. United States Geological Survey, report 2017-5040. Boulder, Colorado.

NOAA, 2018. https://www.ncdc.noaa.gov/billions/events.pdf.

Rhein, M., Rintoul, S.R., Aoki, S., Campos, E., Chambers, D., Feely, R.A., Gulev, S., Johnson, S.A., Kostianoy, A., Mauritzen, C., Roemmich, D., Talley, L.D., Wang, F., 2013. In: Stocker, T.F., Qin, D., Plattner, G.K., Tignor, M., Allen, S.K., Boschung, J., Nauels, A.,

Xia, Y., Bex, V., Midgley, P.M. (Eds.), The physical basis. Contribution of working group 1 to the fifth assessment of the intergovernmental panel on climate change. Cambridge University Press, U.K. and New York.

Taylor, P.C., Maslowski, W., Perlwitz, J., Wuebbles, D.J., 2017. Arctic changes and their effects on Alaska and the rest of the United States. In: Wuebbles, D.J. (Ed.), Climate Science Special Report: Fourth National Climate Assessment, v. 1. U.S. Global Change Research Program, Washington, DC.

USGCRP, 2018. In: Reidmiller, D.R., Avery, C.W., Easterling, D.R., Kunkel, K.E., Lewis, K.L.M., Maycock, T.K., Stewart, B.C. (Eds.), Impacts, Risks, and Adaptation in the United States: Fourth National Climate Assessment, vol. 2. U.S. Global Change Research Program, Washington D.C., USA doi:10.7930/NCA4.2018.

USGCRP, 2018a: Impacts, Risks, and Adaptation in the United States: Fourth National Climate Assessment, vol. 2, In: Reidmiller, D. R., Avery, C. W., Easterling, D. R., Kunkel, K. E., Lewis, K.L. M., Maycock, T. K., Stewart, B. C. (eds.). Water, chapter 3. U.S. Global Change Research Program, Washington D.C., USA. https://nca2018.globalchange.gov/chapter/water.

USGCRP, 2018b: Impacts, Risks, and Adaptation in the United States: Fourth National Climate Assessment, vol. 2, In: Reidmiller, D. R., Avery, C. W., Easterling, D. R., Kunkel, K. E., Lewis, K.L. M., Maycock, T. K., Stewart, B. C. (eds.). Energy, chapter 4. U.S. Global Change Research Program, Washington D.C., USA. https://nca2018.globalchange.gov/chapter/energy.

USGCRP, 2018c: Impacts, Risks, and Adaptation in the United States: Fourth National Climate Assessment, vol. 2, In: Reidmiller, D. R., Avery, C. W., Easterling, D. R., Kunkel, K. E., Lewis, K.L. M., Maycock, T. K., Stewart, B. C. (eds.). Land cover, chapter 5. U.S. Global Change Research Program, Washington D.C., USA. https://nca2018.globalchange.gov/chapter/land-changes.

USGCRP, 2018d: Impacts, Risks, and Adaptation in the United States: Fourth National Climate Assessment, vol. 2, In: Reidmiller, D. R., Avery, C. W., Easterling, D. R., Kunkel, K. E., Lewis, K.L. M., Maycock, T. K., Stewart, B. C. (eds.). Forests, chapter 6. U.S. Global Change Research Program, Washington D.C., USA. https://nca2018.globalchange.gov/chapter/forests.

USGCRP, 2018e: Impacts, Risks, and Adaptation in the United States: Fourth National Climate Assessment, vol. 2, In: Reidmiller, D. R., Avery, C. W., Easterling, D. R., Kunkel, K. E., Lewis, K.L. M., Maycock, T. K., Stewart, B. C. (eds.). Coastal Effects, chapter 8. U.S. Global Change Research Program, Washington D.C., USA. https://nca2018.globalchange.gov/chapter/coastal. Accessed, July 2020.

USGCRP, 2018f: Impacts, Risks, and Adaptation in the United States: Fourth National Climate Assessment, vol. 2, In: Reidmiller, D. R., Avery, C. W., Easterling, D. R., Kunkel, K. E., Lewis, K.L. M., Maycock, T. K., Stewart, B. C., (eds.). Oceans, chapter 9. U.S. Global Change Research Program, Washington D.C., USA. https://nca2018.globalchange.gov/chapter/oceans.

USGCRP, 2018g , Impacts, Risks, and Adaptation in the United States: Fourth National Climate Assessment, vol. 2, In: Reidmiller, D. R., Avery, C. W., Easterling, D. R., Kunkel, K. E., Lewis, K.L. M., Maycock, T. K., Stewart, B. C. (eds.). Agriculture, chapter 10. U.S. Global Change Research Program, Washington D.C., USA. https://nca2018.globalchange.gov/chapter/agriculture-rural.

USGCRP, 2018h: Impacts, Risks, and Adaptation in the United States: Fourth National Climate Assessment, vol. 2, In: Reidmiller, D. R., Avery, C. W., Easterling, D. R., Kunkel, K. E., Lewis, K.L. M., Maycock, T. K., Stewart, B. C. (eds.). Cities, chapter 11. U.S. Global Change Research Program, Washington D.C., USA. https://nca2018.globalchange.gov/chapter/built-environments.

USGCRP, 2018i: Impacts, Risks, and Adaptation in the United States: Fourth National Climate Assessment, vol. 2, In: Reidmiller, D.R., Avery, C.W., Easterling, D.R., Kunkel, K.E., Lewis, K.L.M., Maycock, T.K., Stewart, B.C., (eds.). Human Health, chapter 14. U.S. Global Change Research Program, Washington D.C., USA. https://nca2018.globalchange.gov/chapter/health.

USGCRP, 2018j: Impacts, Risks, and Adaptation in the United States: Fourth National Climate Assessment, vol. 2, In: Reidmiller, D.R., Avery, C.W., Easterling, D. R., Kunkel, K. E., Lewis, K.L. M., Maycock, T. K., Stewart, B. C. (eds.). Northeast, chapter 18. U.S. Global Change Research Program, Washington D.C., USA. https://nca2018.globalchange.gov/chapter/northeast.

USGCRP, 2018k: Impacts, Risks, and Adaptation in the United States: Fourth National Climate Assessment, vol. 2, In: Reidmiller, D. R., Avery, C. W., Easterling, D. R., Kunkel, K. E., Lewis, K.L. M., Maycock, T. K., Stewart, B. C. (eds.). Southeast, chapter 19. U.S. Global Change Research Program, Washington D.C., USA. https://nca2018.globalchange.gov/chapter/southeast.

USGCRP, 2018l: Impacts, Risks, and Adaptation in the United States: Fourth National Climate Assessment, vol. 2, In: Reidmiller, D. R., Avery, C. W., Easterling, D. R., Kunkel, K. E., Lewis, K.L. M., Maycock, T. K. and Stewart, B. C. (eds.). Southwest, chapter 25. U.S. Global Change Research Program, Washington D.C., USA. https://nca2018.globalchange.gov/chapter/southwest.

USGCRP, 2018m: Impacts, Risks, and Adaptation in the United States: Fourth National Climate Assessment, vol. 2, In: Reidmiller, D. R., Avery, C. W., Easterling, D. R., Kunkel, K. E., Lewis, K.L. M., Maycock, T. K., Stewart, B. C. (eds.). Alaska, chapter 26. U.S. Global Change Research Program, Washington D.C., USA. https://nca2018.globalchange.gov/chapter/Alaska.

CHAPTER 6

Adaptation

6.1 Introduction

Societies have a long record of adapting to the impacts of weather and climate change (Chapter 4). As mentioned earlier, early farming communities in northern Syria adapted to a cold-dry event at 8.2 ka (Meslolithic time) by switching livestock from pigs (which were poorly adapted to arid conditions) to cattle, and to a greater use of sheep's wool for warmer clothing (van der Plicht et al., 2011). At the same time in western Scotland a hiatus in radiocarbon dates was interpreted as due to the entire population abandoning the region, because they were not able to adapt quickly enough under rapidly deteriorating conditions (Wicks and Mithen, 2014). Climate change also induces adaptation in natural systems, altering ecosystems and habitats such that some species may or may not be able to adapt fast enough, resulting in extinctions or reduced populations (Settele et al., 2014). Warming trends, on the other hand, tend to promote adaptation by migration of species to higher latitudes or to higher elevation (or deeper water).

Before proceeding it is useful to define some terminology commonly used in the adaptation literature. A hazard usually refers to climate-related physical events or trends and their impacts – they may cause loss of life, injury, or negative health effects. Exposure refers to a wide range of subjects (people, livelihoods, infrastructure, eco-services, resources, and other assets) in places and settings that could be adversely affected. Vulnerability includes the susceptibility to harm and the lack of capacity to cope and adapt. The economically disadvantaged, the elderly, and the young are thought to be vulnerable populations in the face of climate change. Risk refers to the probability of occurrence of hazardous events multiplied by the impacts if such events occurred. Chapter 5 discussed the impacts from climate change in some key sectors and regions of the US risk results from the interaction of hazards, exposure, and vulnerability (Fig. 6.1).

Losses due to climate change in the US by mid-century could reach hundreds of billions of dollars (Bloomberg et al., 2014). Adaptation refers to actions taken at the individual, community, regional, and national levels to reduce risk from current climate change and projected changes. Adaptation

Climate Change in the Anthropocene.
DOI: https://doi.org/10.1016/B978-0-12-820308-8.00008-8

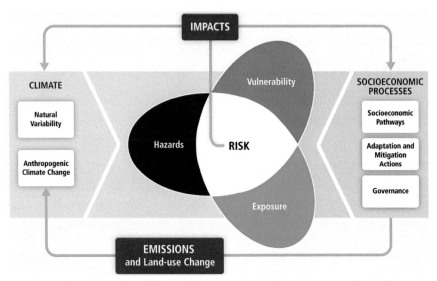

Figure 6.1 Diagram showing risk at the intersection of exposure, hazards, and vulnerability. Impacts are shown as due to socioeconomic processes and climate change, both natural variability and anthropogenic change. *(Source: Summary for Policymakers, IPCC, 2014, Working Group II, fig. SPM.1.)*

is a form of risk management in both the near-term (i.e. by mid-century) and long-term (end of century). These time scales are longer than normally planned for by organizations and governments, thereby hindering implementation of adaptation plans. As outlined in Chapter 3, model projections in the near-term are less sensitive to current emissions as the climate response is already "baked in" (i.e., the climate is not yet in equilibrium with past and current radiative forcings). Long-term projections, on the other hand, depend on the magnitude of the reduction of emissions over coming decades (i.e., which RCP trajectory will likely be followed). Since these paths are challenging to predict, this gives rise to large uncertainty in future climate scenarios (see SSPs below). Under both long-term and short-term time scales, climate impacts will continue to become more extreme unless reductions in emissions are made.

Adaptation can yield benefits in excess of their cost in terms of economic benefits, ecological and health benefits, and human security. In other words, adaptation requires substantial up-front investments but may result in long-term savings over time. The lack of implementation of adaptation plans in the US and elsewhere, especially in developing countries, is likely due to the high initial costs. In Chapter 5, it was concluded that, for example, investment in coastal infrastructure was insufficient to guard against sea level rise, coastal storms, flood control, and coastal protection (USGCRP, 2018, chap. 28). The latter study also concluded that while adaptation plans in the US were well advanced, on-the-ground implementation as of 2018 was not common. Five stages of adaptation are commonly recognized as follows (USGCRP, 2018, chap. 28):

- Awareness
- Assessment
- Planning
- Implementation
- Monitoring and evaluation

This process is not a one-off event but is an ongoing cyclic process involving reassessment, learning and response, sometimes referred to as iterative risk management (USGCRP, 2018, Chapter 2). The overall goal is to reduce exposure, vulnerability, and increase adaptive capacity. Reducing exposure entails reducing people or assets to adverse climate impacts. Reducing vulnerability (sometimes also termed sensitivity) involves lowering the degree to which people or assets are affected by climate impacts. Increasing adaptive capacity entails increasing the ability of systems (natural and manmade) to respond to climate impacts.

Adaptation actions can be taken by individuals, business, government, and society in general on many different scales. Individuals, for example, can obviously guard against flooding by not building in a flood zone, elevating buildings, elevating the electric panel, construct barriers to flooding, and keeping an adequate supply of water, food, and candles. Multiple benefits can arise from adaptation actions. For example, restoring wetlands can provide habitat for fish as well as protecting the coast against storms. Insulating buildings can reduce energy costs and protect against temperature extremes.

Analysis of developed countries' adaptation plans indicates that a common assumption is that future climate will be similar to current and past climate variability, but this assumption is likely incorrect (USGRP, 2018; Wuebbles et al., 2017). Increases in carbon emissions will increase the frequency, intensity, and duration of climate extreme events; in addition, because of the slow response of parts of the climate system (namely the ocean), the climate system has not yet fully equilibrated with emissions that have already occurred. Therefore, projected future climates will be more extreme and more variable than past and present climates (Wuebbles et al., 2017). Adaptation is discussed below under two headings – adaptation needs and adaptation options.

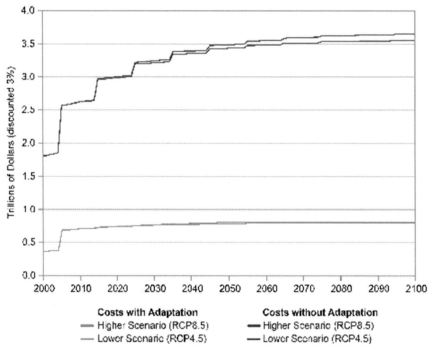

Figure 6.2 Costs with adaptation (lower curves) and without adaptation (upper curves) for two different RCP scenarios. *(Source: USGCRP, vol. 2, 2018, Fig. 8.1.)*

6.1.1 Adaptation needs

Adaptation requires information on risks on specific events occurring and vulnerabilities of populations and groups in order to identify needs to reduce risk. This is best done by engaging people from different backgrounds, experience, and knowledge, and sharing of this knowledge improves adaptation outcomes. The successful implementation of adaptation also depends on the availability of information, access to technology, and funding. For example, the widespread use of mobile phones and the Internet open up possibilities for better adaptation information sharing. In several international climate accords (e.g. the Paris Accords, Chapter 7), developed countries have pledged aid to developing countries to promote adaptation to climate change, but to date much of this funding has not been forthcoming (Fig. 6.2; Chambwera et al., 2014).

6.1.2 Adaptation options

Adaptation options are often addressed under three broad categories: structural/physical, social, and institutional. The structural/physical category can be sub-divided into engineered, technological, ecosystem based, and services. The second category, social, can be subdivided into educational, informational, and behavioral. The third category, institutional, can be subdivided into economic, government policy, and laws and regulations. Some adaptive measures cut across several categories and sectors (Table 6.1).

There can be a gap between adaption needs and effectiveness and this is referred to as the adaptation deficit. An additional useful concept is that of mainstreaming. This refers to integrating climate adaptation into existing structures, procedures and sectors. It works best with governments and organizations that already have well-developed risk management procedures, such as disaster management organizations (e.g., FEMA in the US), the military (e.g., the U.S. Army Corps of Engineers), and the capital investment sector. Below are some examples of the adaptations being undertaken in some US sectors and in some US regions (USGCPR, 2018).

6.1.3 Energy

Superstorm Sandy in 2012 and Hurricanes Harvey, Irma, and Maria in 2017 substantially damaged the energy infrastructure in their respective target areas and provided a learning opportunity on how to rebuild and enhance resilience against future storms. In Puerto Rico especially, the entire electrical grid was destroyed by hurricane Maria partly because it was already poorly maintained. Examples of actions that can be undertaken (requiring major investment in infrastructure) are as follows:

- Elevating equipment, building floodwalls and storm gates, use of smart switches, installing cooling and ventilation systems in case of extreme heat, redesigning electrical circuits.
- Deploying back-up generators, relocating assets, use of microgrids isolated from each other that operate independently.
- Providing alternative cooling systems for power plants during drought.
- Adding redundancy to increase systems resilience to disruption.

6.1.4 Freshwater

As water demand meets supply (e.g., Fig. 6.3) it is a challenge to manage and plan for future climate conditions outside the range of the past climate. Some strategies that may help in management during times of increased variability of water supply include:

- Innovation in management and planning and in infrastructure design.

Table 6.1 Examples of adaptation options.

Category		Examples
Structural/physical	Engineered	Sea walls, levees, coastal protection, water storage, sewage works, improved drainage, building codes, road infrastructure, electric grid, waste water management.
	Technological	New crop and animal varieties, genetic techniques, efficient irrigation, water and food storage, hazard mapping and monitoring, insulation, early warning systems.
	Ecosystems	Wetland and floodplain conservation, reforestation, prescribed burning, fisheries management, land use management.
	Services	Social safety nets, food banks and food distribution, municipal services, vaccination programs, emergency health services, emergency shelters.
Social	Educational	Raising awareness, extensional resources, gender equity in education, conferences and research networks.
	Informational	Hazard and vulnerability mapping, early warning systems, remote sensing monitoring, climate and forecast services.
	Behavioral	Evacuation planning, livelihood diversification, changing agriculture practices, use of societal networks.
Institutional	Economic	Taxes and subsides, insurance, water tariffs, payment for ecoservices, microfinance.
	Government policies	Land zoning and building standards, insurance purchasing, fishing quotas, land rights, water management, coastal management.
	Laws and regulations	National, regional, and local plans, urban upgrades, disaster plans, city development plans.

Source with permission: Adapted from Noble et al., 2014, Table 14.1.

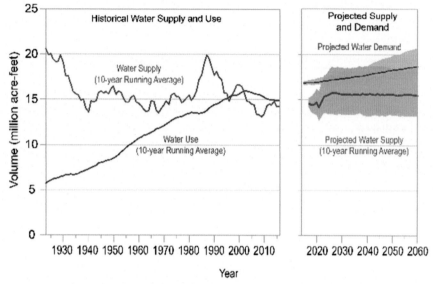

Figure 6.3 Historical water supply and use for the Colorado river basin (left panel). Projected supply and demand (right panel). *(Source: USGCRP, vol. 2, 2018, Fig. 3.3.)*

- Adaptive reservoir operations and water release schedules under more flexible operating rules.
- Updating the regulatory, legal, and institutional structures.
- Funding for maintenance of water infrastructure, monitoring of current conditions, and modeling of future hydrologic scenarios.
- In the case of river flooding, increased use of natural or green infrastructure in the floodplain.
- Include input from all stakeholders (e.g., cities, agriculture, industry, and environmentalists).

6.1.5 Forests

A longer wildfire season has increased the number, intensity, and area of burnings, and insect outbreaks and droughts have also resulted in greater tree mortality. Adaptive strategies include:
- Prescribed burning to reduce the fuel load.
- Reduce forest stand density to help prevent insect outbreaks.
- Planting of drought resistant trees.

6.1.6 Coastal zone

Cities, such as Miami, New York, Boston, Los angeles, New Orleans, and San Francisco have plans to reduce vulnerability to sea level rise, storm surges, and high tide flooding. However, the financial resources committed to date are insufficient to adapt or mitigate against these hazards. Actions that can be undertaken include:

- Installation of flood walls, city storm water pipes, tide gates, pumping stations.
- Conservation of ecosystem functions (e.g., erosion control, fish habitat) and improved coastal protection (e.g., green infrastructure).
- Collaboration between private and public sectors helps defray costs.

6.1.7 Oceans

As outlined previously (Chapter 5), the three main problems in the oceans due to global warming are increasing water temperature, acidification, and deoxygenation. The only adaptive strategy in this case is to reduce global emissions of GHGs as much and as soon as possible. Due to the slow response time of the oceans, this is a long-term (end of century) project.

6.1.8 Agriculture

As outlined in Chapter 5, warming temperatures and changing precipitation patterns threaten agriculture by decreasing food production, increasing soil erosion and decreasing soil moisture. Increase in wildfires (size and number) decreases forage on rangelands. Increase in temperature also reduces water for irrigation and the intensity of drought. Livestock and some crops are less productive under higher temperatures. Adaptive strategies to reduce vulnerability to these threats include:

- Better agricultural practices to restore soil structure (e.g., winter cover crops; no tillage planting).
- Genetically modified higher yield stress-tolerant crops.
- Improved design for cooling of confined animal housing.
- Improved irrigation technology.
- Heat tolerant livestock and increase in shade.

6.1.9 Cities

Current infrastructure design (buildings, roads, bridges, electric grids, water transport and treatment, power production, etc.) does not take future climate change into account and this infrastructure is nearing the end of its useful life span in many cities. Urbanization affects air, water, and soil quality and large impervious areas promote flooding, putting urban ecosystems under stress. The heat island effect, due to large areas of concrete and asphalt and less

Table 6.2 Projected percent increase of very hot days over the next century.[a]

Timeframe	2016–2045	2036–2065	2070–2099
Phoenix, AZ (>110°F)	32% (37%)	121% (215%)	215% (531%)
Fort Collins, CO (>90°F)	60% (80%)	100% (200%)	200% (600%)
Dubuque IA (>90°F)	48% (127%)	139% (298%)	218% (695%)
Pittsburg, PA (>90°F)	94% (108%)	247% (344%)	316% (872)
Charleston, SC (95°F)	33% (44%)	133% (233%)	233% (677%)

[a] Relative to 1976–2005. *Source:* adapted from fig. 11.2, USGCRP, 2018. For RCP 4.5; (RCP8.5).

shade, together with rising temperature means more extreme temperatures are expected (Table 6.2). Several studies have shown that within cities, poorer neighborhoods have a greater heat island effect often due to less shade and less open space (e.g., parks).

Urban social inequality is evident in disparities in per capita income, exposure to violence, environmental hazards, transportation and services, so that different segments of the urban population have different exposure and vulnerability to climate change. Depending on their geography (coastal or inland), cities also have different vulnerabilities to future climate change. Adaptive measures include the following:

- Widespread participation in adaptive decisions (school districts, local government, private sector, utility providers, nonprofit organizations).
- Mainstreaming adaptive measures into existing structures and organizations.
- Increasing green space and tree canopy to help manage storm water runoff and provide shade from extreme heat.
- Avoid development in flood prone areas of the city.
- Cleaner vehicles (natural gas or electric) to reduce city air pollution and reduce GHGs emissions (cities account for 80% of US emissions).
- High albedo roofs to reflect heat reduce building cooling costs.

Some specific examples for various urban areas are outlined below (Vogel et al., 2016):

- El Paso, TX. The El Paso water utility focused on population growth and drought. Together with the US military base at Fort Bliss, in an unusual alliance, they built the Kay Hutchinson desalination plant, which takes brackish water and makes it drinkable.
- Fort Collins, CO. In response to severe drought, the Fort Collins water utility updated its water management policy to (1) put in place regulatory measures to reduce water use quickly during drought and (2) reduce water consumption through conservation.

- Norfolk, VA. Norfolk passed new ordinances after severe flooding. The new ordinances include requiring new structures be built 3 feet above the 100 year floodplain.
- Seattle WA. As an example of mainstreaming, Seattle public utilities integrated climate considerations into 4 levels of their internal planning and operations: (1) organization-wide strategic planning, (2) planning at the water division and sewer division levels, (3) capital investment decision making, and (4) day-to-day operational decisions.
- Mobile, AL. The Nature Conservancy received a grant to rebuild oyster reefs in Mobile Bay after storm inundations damaged coastal ecosystems and fisheries. The restoration helped protect coastal wetlands and the sustainability of the fisheries.
- Tulsa, OK. After several severe flood events it acquired and relocated buildings from the flood zone and turned the flood-prone areas into parks and public areas. Since the 1970s Tulsa has relocated 1000 buildings.

6.1.10 Ecosystem-based adaptation

The above discussions refer largely to human driven adaptation strategies in various sectors of the economy but natural systems will also adapt with or without help from humans. Effects of warming on ecosystems include species migration to more tolerant zones and their place maybe taken by invasive species. Depending on the rate of climate change (which increases with higher RCPs) many species will be unable to migrate fast enough and will decrease in abundance or become extinct; extinction rates are expected to increase with higher RCPs. During the Eocene warming (Chapter 4) boreal forests are thought to have migrated north into the tundra, based on tree pollen studies. Carbon stored in peat lands, permafrost, and forests maybe lost to the atmosphere during deforestation and warming of peat lands and permafrost ecosystems.

Humans depend on ecosystem services for food, freshwater, fiber, health, and security and it is vital that these services be preserved as much as possible. A value of $16 trillion to $54 trillion has been put on these services, globally, but such services are not normally considered part of the economy but are nevertheless essential for our wellbeing (Costanza et al., 1997). Management actions that can help preserve these services have been discussed above and include assistance in species migration, maintenance of biodiversity, reduction in wildfires, and floods and other stressors.

6.1.11 Economics of adaptation

The economics of adaptation are more complex than a cost-benefit analysis – other items to be considered include income distribution, poverty, regional economic activity (e.g., employment), and nonmarket factors, such as water quality and ecosystem functions (Chambwera et al., 2014). Any analysis will be partly economic (cost-benefit) and partly not, also partly quantitative and partly not. The uncertainty regarding future climate change impacts and the associated risks makes any analysis likely to be qualitative with large uncertainty. Adaptation needs the participation of both the private and public actors in its implementation and some examples are listed below:

- Direct capital investment in infrastructure such as dams and bridges (mainly public funding).
- Technological development through research such as new crop varieties (public and private).
- Creation and dissemination of adaptation information (mainly public).
- Human capital investment such as education (public and private).
- Development of adaptation institutions (e.g., new forms of risk insurance; public and private).
- Changes in regulations to facilitate adaptation (e.g., new building codes; mainly public).
- Emergency response procedures (mainly public).

Some adaptation actions have positive ancillary effects (cobenefits) and conversely negative effects (co–costs). For example, a seawall may protect from storm surges, but also protect from tsunamis. However, a co–cost might be damage to nearby ecosystem functions. Reducing energy costs through building insulation may also protect against heat extremes as a co–benefit. Many economists argue that co–benefits and co–costs should be factored into adaptation decision making. This decision making, as mentioned earlier, is a dynamic process and should be flexible over time involving learning and adjustment.

Who pays for adaptation? Developing countries have long argued that since most of the GHGs emissions were produced by developed countries they should accept most of the financial responsibility, and indeed the Paris agreement (UNFCCC, 2015) is framed within this context. Various estimates of the global cost of adaptation (and mitigation) show a range of $70 to $100 billion annually, but the confidence in these estimates is low on account of the large uncertainties involved (Chambwera et al., 2014).

Moreover, existing contributions as of 2020 are only a small fraction of these numbers.

6.1.12 Optimizing adaptation costs

What is the optimal spending on adaptation that will produce the least amount of damage due to climate change? Ideally, an expenditure of $1 on adaptation should produce $1 reduction in climate impact. A smaller investment in adaptation will result in a smaller amount of impact reduction, and larger amounts of expenditure will produce diminishing returns. Fig. 6.4 shows a curve where the vertical axis is the cost of impacts avoided and adaptation costs are shown on the horizontal axis. The optimal adaptation spending occurs where the tangent to the curve has a slope of -45° shown by the dashed lines. Also shown on this diagram is a zone of free adaptation – where the cost of reduction of climate impacts is free; for example, planting seeds at a different date or using irrigating only after sundown. Bringing climate change impacts to zero may be impossible due to technological limitations and the cost may also be prohibitive (Chambwera et al., 2014).

6.1.13 Shared socioeconomic pathways

In Chapter 3, four representative concentration pathways (RCPs) were introduced that span a wide range of alternative radiative forcings and these pathways are commonly used to model plausible future emission and temperature scenarios out to 2100 (see Table 3.1). Which pathway future worlds will follow is unclear – it will likely depend on many variables, including demography (population growth and age distribution), economic factors (e.g., poverty, GDP), and technological growth (especially in the energy sector), as well as political, governance, and social factors among other variables, most of which are difficult to predict.

A group of climate-change scientists introduced the concept of shared socioeconomic pathways (SSPs), which provide a framework for the climate change communities across a broad range of disciplines (including those in working groups II and III of the IPCC reports). Using SSPs as a common starting point with a common set of assumptions future world scenarios are explored (Kriegler et al., 2012; O'Neil et al., 2014; 2016). SSPs are more detailed than RCPs and include more factors, such as energy use and air pollution, urbanization and economic drivers and land-use change in computer climate models. In the Coupled Model Intercomparison Project

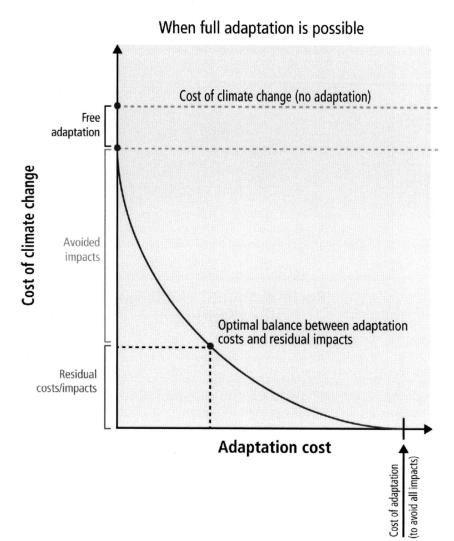

Figure 6.4 Optimizing economics of adaptation (see text). *(Source: Chambwera et al., 2014, Fig. 17.2.)*

Phase 5 (CMIP5; see Chapter 3), RCPs were used to model different future scenarios. The subsequent generation of computer models (CMIP6; Eyring et al., 2016) use SSPs to model plausible future society scenarios.

SSPs are intended to apply globally or to large world regions – they are intended as reference starting points so they assume no climate change policies, other than those presently in force, nor do they assume any climate change (O'Neill et al., 2014). Incorporation of these latter factors was

Figure 6.5 Five standard SSPs. See text. *(Source: O'Neill et al., 2014, Fig. 1.)*

intended for future researchers to develop and evaluate (for example, Riahi et al., 2017). The SSPs are defined in terms of future societal challenges to adaptation and to mitigation (mitigation is discussed in Chapter 7). Fig. 6.5 shows five SSPs on a plot of challenges to adaptation (x axis) versus challenges to mitigation (y axis). SSP1 is characterized by low challenges to adaptation and also low challenges to mitigation. SSP3 is characterized by high challenges to both adaptation and mitigation, and SSP2 is intermediate between these two. SSP4 is characterized by high challenges for adaptation but low challenges for mitigation, and lastly SSP5 is characterized by high challenges to mitigation but low challenges to adaptation. Storylines or narratives for each of these hypothetical future worlds is outlined below (O'Neill et al., 2014; Riahi et al., 2017):

SSP1: Characterized by sustainable development, rapid technological change, lower societal inequalities, and environmentally friendly processes, including lower carbon energy sources and high agricultural productivity. Consumption is oriented toward lower growth and lower resource energy intensity.

SSP2: Is intermediate between SSP1 and SSP3. Social, economic, and technological trends do not shift markedly from historical patterns. Global population growth is moderate and levels off in the second half of the century. Also called the middle of the road SSP.

SSP3: Emissions are high due to moderate economic growth and rapid population growth and slow technological change in the energy sector, making mitigation difficult. Societal inequality is high leaving many people vulnerable to climate change; investment in human capital is low and institutional development is lacking. Regional rivalry leads to poor trade flow and adaptive capacity is low. There is increased nationalism with a focus on domestic issues and concerns about security. Investments in education and technological developments decline. Population growth is low in developed areas but high in less developed regions. Also referred to as the regional rivalry SSP.

SSP4: This is referred to as a mixed world, with regions with high technological development in the energy sector leading to low emissions and high mitigation capacity and other parts with poor development, where inequality remains high and economies are isolated so that adaptive capacity is low. Low technology and high technology societies develop and increased disparity in economic and political opportunity lead to increasing inequality. Social cohesion degrades and unrest and conflict become more common. Referred to as the inequality SSP.

SSP5: Energy demand is high, based on carbon intensive fuels and investment in renewable technology is low leaving few options for mitigation. Nevertheless, economic development is high, leading to less inequality and investment in human capacity is good and institutions are strong. Slower population growth leads to a less vulnerable world able to better adapt to changing climate impacts. Energy and resource intensive lifestyles dominate and investment in health and education is strong. Environmental problems like air pollution are successfully managed and geo-engineering is considered if necessary.

Each of these SSPs can be associated with a specific RCP so that, for example, a society that depends heavily on fossil fuels (e.g., SSP5) can be associated with RCP6.5 and one that is sustainable (e.g., SSP1) can be associated with RCP2.6. These pairings of SSPs and RCPs are denoted in shorthand as SPP5-6.5 and SSP1-2.6, respectively. This has the advantage that results from computer models such as CMIP5 (which uses mainly RCPs as input) and CMIP6 which uses mainly SSPs, can be compared to each other and results from the fifth assessment report IPCC-AR5 (2013) and the sixth report IPCC-AR6 (see Chapter 10) can also be compared to each other. In addition, results across all three working groups

Table 6.3 Characteristics of selected countries.

Country	Population	Energy prod. (Mtoe)	Electric consum. (Twh)	Emissions CO$_2$ Mt
China	1.4 b	2449 (188%)	6302 (987%)	9258 (343%)
USA	328 m	2177 (32%)	4235 (45%)	4896 (2%)
India	1.3 b	554 (98%)	1269 (433%)	2162 (308%)
Nigeria	191 m	249 (71%)	27 (146%)	86 (207%)
France	67 m	135 (21%)	479 (38%)	293 (-15%)

b, billion; m, million; Mt, megatonnes, Mtoe, million barrels oil equivalent; Twh, Tera watts per hour. In parentheses, percent change since 1990. *Source with permission:* IEA, 2018.

(I, II, and III) of these IPCC reports can work from a common set of assumptions.

6.1.14 Current shared socioeconomic pathways

It is interesting to attempt to apply the SSP concept to the current situation in our world today (in the year 2020); this is more an interesting exercise rather than a scientific endeavor. Table 6.3 shows some salient statistics for five countries in order of increasing energy production and electricity consumption and carbon dioxide emissions. In parentheses are also shown the per cent increase change in these statistics since 1990; the populations of these countries are also shown.

In the case of China (population 1.4 billion, and not a democracy), energy production has increased almost 200% since 1990, electricity consumption has increased almost 1000 per cent and carbon emissions have increased 343 per cent since 1990. Pollution in china's major cities is high due to burning of coal. China is co-operating with other nations (mainly by providing loans to developing countries to build infrastructure, including coal power plants) but at the time of writing (2020), rivalry with the US was on the rise (The Economist, 2020). Given these circumstances, China apparently belongs in SSP3, where rivalry is dominant and challenges to mitigation and adaptation are high. India with a similar population as China and with a similar energy profile, is also a rival of its neighbor Pakistan and so also fits into SSP3.

In the case of United States it depends on which administration we focus upon. The Obama administration (2009–2017) promoted renewable energy and tax breaks for electric automobiles and CO$_2$ emissions were only 2% higher than 1990 due largely to a switch from coal to natural gas (obtained using new technology in the energy sector, namely fracking). This would place this administration in SSP1, on a sustainable path. On the other hand, the Trump administration (2016–2020) has reversed many environmentally friendly regulations and favors the coal and fossil fuel industries. This would place this administration in SSP5, where challenges to mitigation are high.

In the case of France, its major dependence on nuclear energy provides for low emissions (Table 6.3) and it continues on a sustainable path (i.e., SSP1), provided serious nuclear accidents can be avoided. Nigeria is a developing country trying to increase its energy production for a rapidly growing population – for now it might fit into SSP2 ("middle of the road"). A major advantage and benefit of the SSP concept is that researchers across a large range of disciplines are employed in collaboration, which was largely absent in the RCP description. Predicting which SSP pathway future worlds will fit into, out to the end of the century, let alone the next decade, is no doubt a major challenge.

References

Bloomberg, M., Paulson, H., Steyer, T.F., 2014. Risky business: the economic risks of climate change in the United States. https://riskybusiness.org. Accessed March 2020.

Chambwera, M., Heal, G., Dubeux, C., Hallegatte, L., Leclerc, A., Markandya, B.A., McCarl, R., Neumann, J.E., 2014. In: Field, C.B. (Ed.), Contribution of Working Group II to the Fifth Assessment Report of the Intergovernmental Panel on Climate Change. Cambridge University Press, Cambridge, UK.

Chambwera, M., Heal, G., Dubeux, C., Hallegatte, S., Leclerc, L., Markandya, A., McCarl, B.A., Mechler, R., et al., 2014. Chapter 17 - Economics of adaptation. In: Climate Change 2014: Impacts, Adaptation, and Vulnerability. Part A: Global and Sectoral Aspects. Contribution of Working Group II to the Fifth Assessment Report of the IPCC. Cambridge University Press.

Costanza, R., d'Arge, R., deGroot, R., Farber, S., Grasso, M., Hannon, B., et al., 1997. The value of the world's ecosystem services and natural capital. Nature 387, 253–260.

Eyring, V., Bony, S., Meehl, G.A., Senior, C.A., Stevens, B., Stouffer, R.J., Taylor, K.E., 2016. Overview of the coupled model intercomparison project phase 6 (CMIP6) experimental design and organization. Geosci. Model Dev. 9, 1937–1958.

IEA, 2018. International energy agency. https://www.iea.org/countries. Accessed March 2020.

IPCC, 2014. Summary for policymakers. In: Field, C.B., Barros, V.R., Dokken, D.J., Mach, K.J., et al. (Eds.), Climate Change 2014: Impacts, Adaptation and Vulnerability. Part A: Global and Sectoral Aspects. Cambridge University Press, Cambridge, UK.

Krieglar, E., O'Neill, B.C., Hallegatte, S., Kram, T., Lempert, R.J., Moss, R., Wilbanks, T., 2012. The need for and use of socio-economic scenarios for climate change analysis: a new approach based on shared socio-economic pathways. Global Environ. Change 22, 807–822.

Noble, I.R., Huq, S., Anokhin, Y.A., Carmin, J., Goudou, D., Lansigan, F.P., Osman-Elasha, B., Villamizar, A., 2014. Adaptation needs and options. In: Field, C.B (Ed.), Climate change 2014 - Impacts, Adaptation and Vulnerability. Part A: Global and Sectoral Aspects. Cambridge University Press, Cambridge, UK.

O'Neill, B.C., Kriegler, E., Riahi, K., Ebi, K.L., Hallegatte, S., Carter, T.R., 2014. A new scenario framework for climate change research: the concept of shared socio-economic pathways. Clim. Change 122, 387–400.

O'Neill, B.C., Tebaldi, C., van Vuuren, D., Eyring, V., Friedlingstein, P., Hurtt, G., et al., 2016. The Scenario Model Intercomparison Project (ScenarioMIP) for CMIP6. Geosci. Model Dev. 9, 3461–3482.

Riahi, K., van Vuuren, D., Kriegler, E., Edmonds, J., O'Neill, B.C., Fujimori, S., 2017. The shared socio-economic pathways and their energy, land use and greenhouse gas emissions implications: an overview. Glob. Environ. Change 42, 153–168.

Settele, J., Scholes, R., Betts, R., Bunn, S., Leadly, P., Nepstad, D., Overpeck, J.T., Taboada, M.A., et al., 2014. In: Field, C.B., et al. (Eds.), Contribution of Working Group II to the Fifth Assessment Report of the Intergovernmental Panel on Climate Change. Cambridge University Press, Cambridge, UK.

The Economist, 2020, March 21st –27th. Sino-American rivalry, p. 37. London.

UNFCCC, 2015. Paris Agreement. United Nations Framework Convention on Climate Change. https://unfccc.int/files/essential_backfround/convention/application/pdf. Accessed March 2020.

USGCRP, 2018. In: Reidmiller, D.R., Avery, C.W., Easterling, D.R., Kunkel, K.E., Lewis, K.L.M., Maycock, T.K., Stewart, B.C. (Eds.), Impacts, Risks, and Adaptation in the United States: Fourth National Climate Assessment, vol. 2. U.S. Global Change Research Program,, Washington D.C., USA doi:10.7930/NCA4.2018.

Van der Plicht, J., Akkermans, P., Nieuwenhuyse, O., Kaneda, A., Russell, A., 2011. Tell Sabi Abyad, Syria: radiocarbon chronology, cultural change and the 8.2 ka event. Radiocarbon, 53, 229–243.

Vogel, J., Carney, K.M., Smith, J.B., Herrick, C., Stults, M., O'Grady, S., St. Juliana, A., Hosterman, H., Giangola, L., 2016. Adaptation: the state of practice in U. S. communities. The KRESGE Foundation and Abt Associates.

Wicks, K., Mithen, S., 2014. The impact of the abrupt 8.2 ka event on Mesolithic population of western Scotland: a Bayesian chronological analysis using activity events as a population proxy. J. Arch. Sci. 45, 240–269.

Wuebbles, D.J., Easterling, D.R., Hayhoe, K., Knutson, T., Kopp, R.E., Kossin, J.P., Kunkel, K.E., LeGrande, A.N., Mears, C., Sweet, W.V., Taylor, P.C., Vose, R.S., Wehner, M.F., 2017. Our globally changing climate. In: Wuebbles, D.J., Fahey, D.W., Hibbard, K.A., Dokken, D.J., Stewart, B.C., Maycock, T.K. (Eds.), Climate Science Special Report. The United States Government's Fourth National Climate Assessment, vol. 1. Washington DC, USA.

CHAPTER 7

Mitigation

7.1 Introduction

Mitigation is the reduction of greenhouse gas emissions or enhancement of their sinks by humans. As part of a broader framework mitigation and adaptation go hand in hand (Chapter 6). The mandate given to working Group III of the fifth Intergovernmental Panel on Climate Change included the following: to be explicit about mitigation options; to be explicit about their costs and risks and to be explicit about concepts and methods for evaluation of alternative policies (Edenhofer et al., 2014). Because climate change is a global problem international cooperation is required for mitigation and this cooperation should include equitable sharing in the effort. This can perhaps be best accomplished through international agreements, although such agreements have met with mixed results so far in terms of meeting their stated goals (e.g., UNFCCC, 2015). Mitigation, especially among developing countries, can lead to tensions between development on one hand and mitigation actions on the other. Developments such as increased energy consumption, infrastructure, or more intensive agriculture can lead to greater emissions and these are often urgent priorities for developing nations. The International Energy Agency (IEA) published the "world's first comprehensive energy roadmap to net zero emissions by 2050" (IEA, 2021). They conclude that nations must move fast to achieve net zero, that the path is narrow, but still achievable. Given that this organization is an advocacy and advisory group for the fossil fuel industry, this publication is a significant event. It is the subject of Chapter 9.

Chapter 2 discussed some of the main GHGs and how their concentration has varied over time and it was pointed out that the GHGs have different radiative forcing values and also different residence times in the atmosphere. For example, methane has a lifespan of about a decade but it is more difficult to specify uniquely the lifespan of carbon dioxide as it has several sources and sinks, but a timescale of centuries or longer is usually associate with CO_2 (Shine et al., 1990). Near-term warming projections (e.g., until mid century) are not dominated by near-term emissions but rather by natural climate variability and current and past emissions, and also by the range in different

Climate Change in the Anthropocene.
DOI: https://doi.org/10.1016/B978-0-12-820308-8.00003-9

computer models. Large reductions in long-lived GHGs will have only a modest effect on surface temperature in the near-term but large reductions in short-lived gases could have a large effect on surface temperature in the near-term (DeAngelo et al., 2017). Long-term projections (out to 2100) will depend on which RCP pathway (Chapter 3) or which SSP trajectory is followed (Chapter 6).

The concept of Global Warming Potential (GWP) was introduced in chapter 2 which measures how strong different GHGs are relative to CO_2 and it is commonly calculated over a 20 year or 100 year timeframe. In the literature the aggregate GHG emissions are commonly referred to as CO_2-equivalents based on Global Warming Potentials with a 100 year time horizon (GWP_{100}). The majority of changes in GHG emission trends are related to economic growth, changes in technology, and or population growth. As outlined below, the corona virus pandemic of 2020 resulted in the largest reduction in emissions of CO_2 in over seventy years.

Cumulative CO_2 emissions have a nearly linear relationship with global mean temperature change making it a key climate indicator (Myhre et al., 2013). Between 1870 and 2015, fossil fuel burning and deforestation have emitted about 2050 $GtCO_2$. It is estimated in order to avoid a 2°C or less increase in the preindustrial temperature (with a 66% probability) no more than 2890Gt CO_2 can be emitted (this estimate includes non-CO_2 GHGs, aerosols, and black soot) (Collins et al., 2013; DeAngelo et al., 2017). This means only 840 $GtCO_2$ can be emitted in order to avoid the 2°C warming limit in the Paris Accords (UNFCCC, 2015). If it is assumed we follow RCP4.5, this limit will be reached by 2037, which at the time of writing is only 17 years away. These considerations indicate that actions to mitigate GHG emissions need to begin as soon as possible if the Paris agreement temperature goal is to be met (Fig. 7.1). It is generally agreed that if the temperature increase is to be stabilized at some level in the future, the net emissions will have to reach zero at some point in the future (DeAngelo et al., 2017).

Countries that announced target reduction in emissions (formally called Intended Nationally Determined Contributions) for 2025 or 2030 leading up to the Paris agreement of 2015 are likely insufficient to meet the 2°C or less goal, unless emissions are net negative (−1.4 GtC/yr.) by 2085 (Sanderson et al., 2016). The fourth U.S. National Climate Assessment reached a similar conclusion (DeAngelo et al., 2017). This would require some form of carbon dioxide removal at scale, which has yet to be demonstrated.

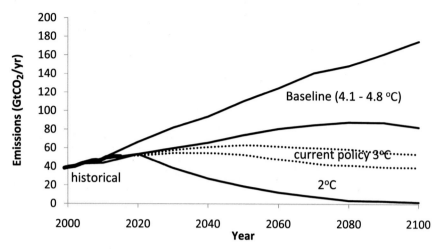

Figure 7.1 Carbon dioxide global emissions (GtCO$_2$/yr) projected to the end of century showing projected temperature increases for three different scenarios. The baseline projection (4.1°C–4.8°C above preindustrial temperature) assumes no new policies and is based on IPCC5-working group III. The current policy (2.8°C–3.2°C) is based on current climate emissions policies around the world. The third scenario (2°C) shows the emissions required to be consistent with a 2°C increase showing close to zero net emissions by 2080. Part of the historical record is also shown. *(Source: with permission based on data from Climateactiontracker.org.)*

7.2 GHG emission trends

Fig. 7.2 shows carbon dioxide emissions per annum for the period 1971–2017. The global (uppermost) curve for 2017 indicates 32.8 GtCO$_2$ emissions (due to fossil fuel burning and cement production) for that year and over the period 2000–2017 the annual increase was 1.7 percent per annum. The increase in the slope of the global curve after 2000 appears to be largely due to the non–OECD countries (the Organization for Economic Cooperation and Development represents 37 of the worlds' leading economies) and also China. Somewhat surprisingly, the US and OECD (Europe) and India do not contribute significantly to this upturn. This pattern underscores the point that developing countries are vigorously expanding energy production while developed countries already have mature energy sectors. India is still in the process of increasing its energy supply, mainly by building coal-fired power plants, and will probably show increased emissions, similar to that of China, in the near future.

The global corona virus (COVID-19) pandemic of 2020 resulted in a temporary negative 175 (-11% to -24%, ±1σ range) decline in CO$_2$ emissions

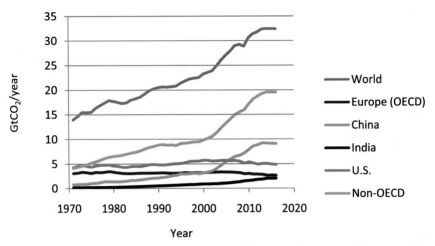

Figure 7.2 Carbon dioxide emissions (GtCO$_2$/yr) for selected regions and countries for the period 1971–2017. Total world emissions are also shown. *(Source with permission: based on data from the International Energy Agency (Annual historical series: IEA, 2020).)*

in April of 2020 (relative to the mean for 2019) but rapidly rebounded to just negative 5% by June 2019 when the global economy began to open up again (LeQuéré et al., 2020). The main contributions to emission reductions were power generation, road transportation, and industry, in that order. The latter authors estimated that the reduction in emissions for all of 2020 to be about 5% (similar to the estimate of the International Energy Agency) and that such reductions would be required every year for several decades to meet the Paris Accord agreement goal of ≤2°C warming over the preindustrial level. Obviously, extreme measures such as shutting down the global economy, as happened for part of 2020, is not an option to mitigate global warming – decarbonization of the energy sector and removal of carbon dioxide from the atmosphere are probably both required (see below). The economic collapse caused by the corona virus pandemic starkly illustrates the challenges we face if global warming is to be kept under control (Box 7.1).

BOX 7.1 Annual CO$_2$ concentration increase in the atmosphere.
We can estimate the approximate increase in concentration in the atmosphere due to global CO$_2$ emissions in a given year. For example, the emissions for 2017 (due to fossil fuel burning and cement production) were 32.8 GtCO$_2$ metric tons (Fig. 7.2). If we divide this figure by the mass of the Earth's atmosphere (5 x 10^{15}

metric tons; Weinstein and Adam, 2008) we get:

$$(32.8 \times 10^9)/(5 \times 10^{15}) = 6.6 \times 10^{-6} = 6.6\,\text{ppm}$$

This figure must be adjusted for the different molecular weights of CO_2 (44 g) and air (29 g) using the factor 29/44 (0.66) to give 4.4 ppm. Since the observed increase at Mauna Loa for the past decade is about 2.5 ppm per annum, it must be assumed there are other sinks for atmospheric carbon dioxide – which are mainly the oceans and the continents. This would imply these sinks accounted for 43% of CO_2 drawdown in 2017. The fifth IPCC-AR5 report (2013) estimated that the oceans alone draw down was about 30% of atmospheric carbon dioxide but the amount depends of several variables related to the oceans' carbon cycle (Rhein et al., 2013). The contribution of the continents depends on land-use changes such as forestation and deforestation and agricultural practices and it is difficult to evaluate, with a large uncertainty (Clarke et al., 2014). The concentration of CO_2 in the atmosphere (420 ppm in 2021 at Mauna Loa) did not decrease due to the pandemic virus due to the relatively short time span of reduced emissions and the long residence time of CO_2 in the atmosphere.

7.3 Emission drivers

A long-standing identity of environmental impact (Ehrlich and Holdren, 1971) is: $I = PAT$ (referred to as $IPAT$), where I refers to environmental impact (e.g., GHG emissions), P refers to population and A refers to affluence (e.g. income per capita), and T refers to technology (e.g., GHG emission intensity). A modified version of $IPAT$ identifies four main drivers of CO_2 global emissions (Raupach et al., 2007) namely, (1) population, (2) per capita GDP (GDP/population), (3) energy intensity of GDP (energy consumption/GDP), and (4) CO_2 intensity of energy (CO_2 emissions/energy). This can be written as:

$$F = P(GP)(EG)(FE) = \text{Pgef } F = PGPEGFE = \text{Pgef} \qquad (7.1)$$

where F is global emissions from fossil fuel combustion and industrial processes, P is population, G is world GDP or gross domestic product, E is global energy consumption (from fossil fuels, nuclear and renewable); g is G/P (income per person), e is E/G (energy use per dollar GDP), and f is F/E (carbon intensity). Upper case letters refer to extensive variables and lower case letters refer to intensive variables (ratios).

Raupach et al. (2007) examined the variables in Eq. 7.1 as global averages over the period 1980–2004. As might be expected, as GDP and population increase, emissions increase over time; on the other hand, energy intensity (e) and carbon intensity (f) decreased over time, but with an upturn in both e and f since 2000, largely attributed by the authors to increased use of coal in power generation. If GDP decreased, as in a recession, e would also increase if energy consumption remained the same; but that is unlikely, so that the two variables (E/G) are not independent and compete against one another. As noted above (see Fig. 7.2) the rate of global emissions increased after the year 2000 and Raupach et al. (2007) attributed this pattern to increasing trends of carbon intensity of energy (i.e. burning of more coal), and increasing trends in energy intensity of GDP (e) in addition to population growth (P). These trends need to be studied carefully if mitigation actions are to be successful (see SSP paths below).

Trends in population and demographic structure are mostly shaped by levels of fertility and mortality, which have universally declined over the past few decades resulting in lower global population projections by the United Nation's Population Division. In 1950, the global fertility rate was five, but is predicted to approach two live births by the end of the century, which is close to the fertility rate for a stable population (i.e., zero growth rate). The estimated global population in 2020 was 7.8 billion and the median variant scenario for 2050 is 9.7 billion (Fig. 7.3). Small changes in fertility rates make for large changes in population over time. If the fertility rates increased by +0.5, the projected population by 2050 is 10.9 billion, and if the fertility rate decreased by -0.5 the projected population peaks at 8.9 billion in 2050 and it decreases thereafter (Fig. 7.3).

Regional trends in the other variables in Eq. 7.1 are somewhat complex but they must be examined if mitigation of global emissions is to be addressed. Raupach et al. (2007) studied four large emitter nations (US, China, India, and Japan), the European Union and the former Soviet Union and other developed and developing nations over the period 1980–2004. The US and the European Union and other developed nations show a pattern similar to the global pattern outlined above. China showed very large increases in GDP and emissions (F) and also a large decrease in carbon intensity (f) and a smaller decrease in energy consumption per GDP (e). The former Soviet Union shows a unique pattern at the time of the break-up, so that the post-1990 GDP decreases precipitously along with emissions (F), and e and f show large increases over 1990 values, a very different pattern from other countries and regions; this pattern begins to reverse itself by 2005 as their economy recovers.

World: Total Population

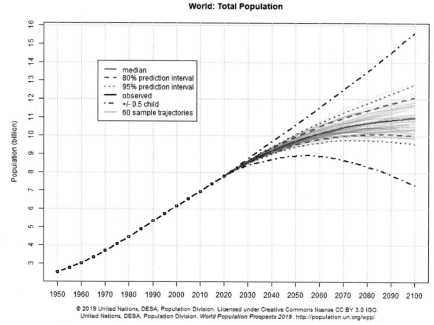

Legend:
- median
- 80% prediction interval
- 95% prediction interval
- observed
- +/- 0.5 child
- 60 sample trajectories

Figure 7.3 World population projections to the end of the century showing median estimate at 2100 of about 10.9 billion and for fertility rates of +0.5 (15 billion) and -0.5 (7.5 billion). *(Source: U.N. world population prospects (2019). http://population.un.org/wpp.)*

7.4 Carbon intensity of energy

Carbon intensity of energy $(f = F/E$; Eq. 7.1) can be measured in grams of carbon per megajoule (gC/MJ) with typical values for various regions being between 15 and 20 with India and China being greater than 20 (largely due to coal burning) with a world mean value of 17 (Raupach et al., 2007). The fossil fuels, coal, oil, and natural gas produce decreasing amounts of carbon dioxide when burned in the ratios: $1, 1/2, 1/3$, respectively. Ignoring the mineral fraction, coal has the approximate chemical composition of the element carbon (C) and petroleum can be approximated by CH_2, and methane by CH_4. When coal is burned it is converted to CO_2 producing 3.6 g (44/12) for every gram burned. When petroleum is burned two reaction occur, one producing water and the other carbon dioxide, so that it produces half as much CO_2 as coal. When methane is burned three reactions occur: two molecules of water are produced and one molecule of carbon dioxide, so that it produces $1/3$ CO_2 compared to coal (see Table 7.1). One obvious route to mitigating emissions is to emphasize renewable sources and natural gas as outlined in the IEA's sustainable Development Scenario (International Energy Agency, 2020) (Box 7.2).

Table 7.1 Carbon dioxide emissions of fossil fuels.

Fossil fuel	Composition	Combustion reactions	Amount of CO_2
Coal	C	1	1
Oil	CH_2	2	1/2
Natural gas	CH_4	3	1/3

BOX 7.2 The Paris Accords.

The Paris agreement on climate change went into effect on December 15, 2015 (UNFCCC, 2015). It is a relatively short document at 25 pages and 27 articles. About 190 countries, including the European Union, signed the agreement including the United States and China, the latter two countries representing more than 40% of global emissions. As soon as President Trump was elected in 2016 he said the United States would withdraw from the agreement – but this withdrawal can only go into effect after 4 years from the time of withdrawal, namely 2020. President Biden re-joined the agreement in 2021. A recurrent theme throughout the agreement is that developed nations should help underdeveloped nations in the transition to lower emissions through technological transfer and financial aid.

The most important and notable feature of the accord is that the magnitude of reduced emissions are up to each individual nation to specify themselves and are therefore voluntary. The major goal of the accord is to limit global warming by the end of the century to less than 2°C over preindustrial levels. Since we already have 1°C warming since preindustrial time (Chapter 1), this means we must limit additional warming to less than 1°C by the end of the century. Additional recurring themes in the accord are that the transition to low emissions should be sustainable and should protect indigenous peoples and other vulnerable populations (e.g., inhabitants of small low lying islands) with an emphasis on social justice, reducing poverty, and also protecting ecosystems. Nations must periodically report to the United Nations on how they are progressing on their emission pledges.

7.5 Sectors

7.5.1 Energy

Fig. 7.4 shows the global energy supply from 1990 to 2015 and not surprisingly, fossil fuels dominate the energy mix. Although wind and solar are at the bottom of the diagram they nevertheless account for about 50% of recently added new electricity capacity, indicating their potential future importance. Although with almost net zero emissions, the nuclear energy

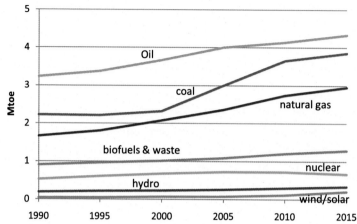

Figure 7.4 Energy production for the period 1990 to 2015 by source in million tonnes of oil equivalent (Mtoe). *(Source with permission: data from International Energy Agency (energy balances, 2019).)*

contribution is nearly flat with some recent downturn; the serious nuclear accident in 2011 at Fukushima, Japan, has resulted in adverse public opinion and less investment in this sector. As noted already, coal shows a steep climb after 2000. Fig. 7.4 together with Fig. 7.1 (CO_2 emissions) can be used as a basis for discussion of climate mitigation in the energy sector.

The United Nations (UN) developed 17 sustainable development goals (SDGs) to be met by 2030 that were adopted by member states in 2015 (United Nations, 2020). Sustainable development is defined as meeting the needs of the present generation without compromising the needs of future generations. The IEA subsequently focused on three of these goals that the energy sector could achieve and the Agency refers to them as their sustainable development scenario (SDS) (IEA, 2020). The three goals are universal access to energy (UN SDG7), reduce severe health impacts from air pollution (UN SDG3) and tackle climate change (UN SDG13). The latter goal is designed to be consistent with the Paris Accords of a less than 2°C temperature increase. The Agency starts with these goals and works backwards to achieve the goals in a realistic and cost-effective way. To achieve these goals the Agency concluded that the CO_2 emissions of 37 billion tonnes in 2018 should drop to 10 billion tones by 2050 and to zero net emissions by 2070. Carbon capture and storage (CCS) would be required to reach these goals, which has not yet been demonstrated on a large enough scale (see below); current CCS amounts to 40 million tonnes per year, equivalent to 0.15% of all emissions. As mentioned above the more recent publication (IEA, 2021) aims to reach net zero emissions by 2050.

Increase in investment, mainly in the electricity sector, would need to be \$45 billion per year, about 2% of the investment of the entire sector investment. This investment would need, however, to be directed mainly at the developing world; the global pandemic virus of 2020, which caused economic global collapse, has slowed progress toward such a goal.

The energy sector accounts for 35% of all emissions and as noted already showed a rate increase after 2000 (the annual increase was 1.7% for 1990–2010 and 3.1% for 2000–2010) Bruckner et al. (2014). Progress in emissions reduction in electricity generation shows greater promise compared to the transportation and industrial sectors. Steps that can be realistically taken toward emissions mitigation in the energy sector are listed below (Bruckner et al., 2014):

- Reduction of fugitive emissions (CO_2 and CH_4) associated with fuel extraction, refining, transport, and energy distribution.
- Fossil fuel switching (e.g., coal to natural gas) although this is insufficient by itself to keep CO_2 concentration at reasonable levels.
- Renewable energy. Recent reduction in cost (mainly wind and solar) have made them competitive with fossil fuels. Taxes on carbon emissions would encourage further investment by electricity generators in this area. Cobenefits include less air pollution (improved health), local jobs, and increased access to energy.
- Nuclear energy. There are several barriers, however, to a greater role for nuclear in the energy mix, including adverse public opinion, operational, financial, and regulatory risks, unresolved waste storage problems, and uranium mining risks. New fuel cycles and reactor technologies address some of these problems but are still in the early development stage.
- Carbon capture and storage (CCS). Oil companies regularly inject CO_2 into depleted oil formations, a process known as enhanced oil recovery, so that this technology is well known. So far only 25 million tones have been permanently stored in the power and industry sectors as of 2019 (GCCSI, 2020). A modified version of CCS combines burning of biofuels (which is a carbon neutral process) together with capture and storage (BECCS), resulting in net negative emissions; in 2019 only one such plant existed (see Chapter 6). Large-scale deployment causes concern for long-term stability of CO_2 storage and transport. A large natural CO_2 release from a volcanic lake occurred in 1986 (lake Nyos in the Cameroon) suffocating 1700 people and 3500 livestock.

7.5.2 Transport

This sector includes light and heavy-duty vehicles, aircraft, shipping/boating, and rail. Greenhouse gas emissions from the transport sector have more than doubled since 1970 and around 80% of this has come from road vehicles (Sims et al., 2014). Globally, the transport sector accounts for 24% of all emissions (28% in the US) and is growing at a faster pace than other sectors, especially in the developing world and economies in transition where economic and population growth are higher. Mitigation of emissions in this sector will be challenging unless emissions can be decoupled from GDP through policy and or regulation. Barriers to mitigation in this sector include the high cost of low-emission transport systems (e.g., rail) and the slow turnover of stock and infrastructure. The cost of emission mitigation is measured as dollars invested per tonne CO_2 mitigated ($/tCO_2$).

Emissions in this sector are more complex than getting from A to B. Total emissions are related to several factors including: system infrastructure (roads, rail, airports) and urban form (sprawling or compact cities); fuel carbon intensity (e.g., gasoline, natural gas, biofuels, diesel, hydrogen) is measured in tonnes of CO_2 equivalent per megajoule (tCO_2/MJ); energy intensity (walking, bicycle, light rail, train, aircraft) is measured as megajoules per passenger-kilometer (MJ/p-km), or in the case of freight, megajoules per tonne-kilometer (MJ/t-km). The amount of activity is also clearly important – the number and length of journeys and journey avoidance (combined journeys or video conferences). Behavioral choice also plays a role – involving factors such as speed, convenience, and cost.

Some progress has been made in the following areas (Sims et al., 2014):
- Renewed interest in compressed natural gas as a fuel for road vehicles.
- Increase in number of electric vehicles (e.g., China) but from a low base.
- Improved fuel economy and GHG emission standards.
- Improved urban planning and development for pedestrians and bicycles, buses, and light rail.
- Reduced carbon intensity of operations by freight companies.
- There is some evidence that the number of light vehicles (e.g., autos) in the developed world (OECD countries) may plateau soon.
- Cobenefits of mitigation in this sector include increased access, improved health (reduced air-pollution), and safety.

The fifth IPCC report (Working Group III) concluded that a range of strong mutually-supportive policies will be required to decarbonize the transport sector (Sims et al., 2014).

7.5.3 Industry

Industry involves the conversion of natural resources into material stocks then converted in manufacturing and construction into products (Fischedick et al., 2014). Industrial emissions in 2010 (13 $GtCO_2$) accounted for a total of 28 percent of all emissions, higher than other sectors. Industrial subsectors include iron and steel, cement, mining, chemicals, pulp and paper, aluminum and textiles, and leather.

In a manor somewhat similar to that of Eq. 7.1 above, industrial emissions can be represented in simplified form as follows:

$$G = (GE)GE$$

$$(EM)(MP)(PS) \, EMMPPSS \tag{7.2}$$

where G is the yearly greenhouse emissions of the industrial sector; E is the industrial sector energy consumption, M is the total yearly global production of materials; P is the stock of products created from these materials, and S is the services delivered through the use of these products.

7.5.4 G/E is emissions intensity

This is the ratio of emissions to energy consumption either directly from burning fuel and indirect consumption from chemical reactions (termed process emissions). Energy intensive processes include cement production, chemicals, iron and steel, and aluminum. Most energy consumption uses fossil fuels and a switch to natural gas could reduce emissions. The cement industry in Europe incinerates municipal waste and waste sludge and the European paper industry gets 50% of its energy from biomass. If electricity were decarbonized, reduced emissions could be obtained by replacing boilers with heat pumps. Thermal solar energy can be used for washing, drying, and evaporation but this is not a widespread application. The IEA estimates that carbon capture and storage (CCS) could reduce emissions by 30% by 2050.

7.5.5 E/M is energy intensity

This is the ratio of energy consumption to total production. Energy in industry is used to drive chemical reactions, create heat, and perform mechanical work all of which are subject to thermodynamic limits. It is estimated there is room to increase energy efficiency by up to 25% in particular sectors (Fischedick et al., 2014). Opportunities for greater energy efficiency can be found in: steam systems, heating systems (furnaces, boilers), electric motor systems (pumps, fans, compressors, refrigerators) and electronic control systems can help optimize these systems. Other opportunities include better heat management between hot and cold fluids and gases, capture and use of exhaust heat, and better insulation. Recycling reduces both energy consumption and increases total production, thereby reducing E/M.

7.5.6 M/P is material intensity in product design

This is the ratio of global materials used to make products from these materials. Recycling requires fewer materials (resources) and is widely practiced in the metals industry. Fly ash from coal-fired power plants can be recycled as cement clinker or bricks. Reuse of old materials (e.g., structural steel) provide a profit opportunity. Concrete, however, cannot be recycled, except for aggregates, which itself requires energy for crushing. New lighter materials in autos reduce the amount of steel required, but as cars are getting bigger with more features, this trend is being offset. High labor costs relative to materials also inhibit this trend. Commercial buildings can be built with less steel (up to 50% less) and still meet safety requirements.

7.5.7 P/S is product service intensity

This is the ratio of products to services provided by these products. Lengthening the life of durable goods products also reduces demand for resources (lower emissions), and these activities tie into the concept of sustainable consumption. Reduction of waste (e.g., food and other materials such as scrap metals) also provides a means of increasing the services of a product. From Eq. 7.2 it is clear that reduction in demand for services (S) results in reduced emissions and this is strongly related to behavioral choices and lifestyles, corresponding to less unnecessary consumption in general (i.e., sustainable consumption).

Some other important sub-sectors within industry (cement, chemicals, iron, and steel) are briefly addressed below. In 2016, cement production emitted about 2.8 billion tonnes of CO_2 corresponding to approximately 8% of global emissions. The main reaction in cement production (calcination) is: $CaCO_3 + heat = CaO + CO_2$. About 50% of emissions are due to this reaction so reduction of emissions in cement production is difficult. Most of the remaining emissions are due to heating of calcium carbonate in a kiln. Carbon capture and storage (CCS) could potentially reduce emissions but this not yet widely practiced on a commercial basis.

The chemicals sector produces a wide range of products on a range of scales making data collection and analysis of this sector challenging. There are however a relatively small number of widely used key products, such as ethylene (C_2H_4), ammonia (NH_3), fertilizer, plastics, acids, and synthetic fibers. Emissions arise from use of energy in production and from venting of by-products from chemical processes. Production of light hydrocarbons by steam cracking is one of the most energy intensive processes in the chemical industry and about a 25% decrease in energy efficiency is possible using the best technology in this area. Recycling of plastics saves energy but requires a pure waste stream. Natural gas based ammonia production can reduce emissions by 58% compared to that based on coal, amounting to a reduction of 27 $MtCO_2$ emissions per year.

Steel (made from pig iron) dominates global metal production and CO_2 emissions from this sector were 2.6 $MtCO_2$ in 2019 ($\sim$8 % total emissions). The coal and coke used to make iron in a blast furnace is emissions intensive. Reduction in emissions could be made if low carbon electricity sources are used in electric arc furnaces and newer processes and best technologies are deployed. Energy efficiency is being pursued with improved heat and energy recovery from gases and improved fuel delivery to furnaces.

Fischedick et al. (2014) conclude that a common barrier to mitigation of emissions in industry is lack of information, and initial investment costs, but could be promoted by regulatory approaches, economic incentives, and voluntary actions. These authors also concluded the energy intensity of this sector could be reduced by 25% through widespread upgrading, replacement, and deployment of the best technologies.

7.6 Buildings

Buildings represent a critical part of a low-carbon future and a global challenge for integration into sustainable development (Lucon et al., 2014). Mitigation in this sector, other than cost savings, are accompanied by a diverse set of cobenefits, including energy security, fewer energy subsides, health benefits from cleaner air (inside and outside), environmental benefits, and employment gains. Buildings represent one of the biggest unmet energy needs, especially in the developing world, where slum dwelling in cities is on the increase. It is estimated that 3 billion people rely on highly polluting traditional fuels for cooking and heating, and over 1 billion people worldwide lack electricity. On the other hand, in the developed world, energy use is wasteful and inefficient.

Table 7.2 Typical and best case specific energy consumption (kWh/m²/yr) for buildings.

End use	Climate	Residential		Commercial	
		Advanced	**Typical**	**Advanced**	**Typical**
Heating	Cold	15–30	60–200	15–30	75–250
Heating	Moderate	10–40	40–100	10–30	40–100
Cooling	Hot-dry	0–10	10–20	0–10	20–50
Cooling	Hot-humid	3–15	10–30	15–30	50–150

Source with permission: Adapted from Lucon et al., 2014, Table 9.3.

Buildings are responsible for one third of global final energy consumption and 40 per cent of CO_2 emissions (10 $GtCO_2$ in 2019). In order to avoid locking in energy intensive options for decades, especially in the least developed nations, a shift to electricity, modern fuels, and energy saving solutions (technological and architectural) in addition to energy policies are required. Several factors are driving the increase in building energy use: migration to cities, decrease in household size, increasing levels of wealth, and an increase in the number of appliances in use.

The average residential energy use for the developed world is as follows, for 2010: space heating (32%), cooking (29%), water heating (24%), appliances (9%), lighting (4%), cooling (2%) (Lucon et al., 2014). A useful voluntary residential standard for efficiency use is the Passive House standard which prescribes no more than 15 kWh/m²/yr for heating (assuming 20°C indoors) and annual energy use for appliances of 120 kWh or less. Features of the Passive House standard include very good wall and window insulation, ventilation with heat recovery from exhaust air, and air tightness, and the use of traditional passive design principles; under these conditions air conditioning is not needed in most climates. Solar panels and subsoil heat exchangers are optional. Table 7.2 compares this standard to typical residential energy use in different climates; also shown is data for commercial space. Clearly there is a large potential for mitigation to be gained in this sector. Factors that drive CO_2 emissions from energy use in this sector are: carbon efficiency (switch to low carbon fuels for electricity generation), technological efficiency (energy efficiency of appliances), infrastructure efficiency (building design), and reduction in demand (e.g. lifestyle choices).

7.7 Shared socioeconomic pathways – quantifying the paths

In Chapter 6 we introduced five standard shared socioeconomic pathways (SSPs) which represent possible trajectories for future societies in terms of global or regional socioeconomic development. These pathways were named, from SSP1 to SSP5, as follows: Sustainability, Middle of the Road, Regional Rivalry, Inequality, and Fossil-fuel Development. These pathways face different challenges in terms of mitigation and adaptation with SSP1 facing the lowest challenges and SSP3 facing the highest challenges (see Fig. 6.5). It was also pointed out in Chapter 6 that these SSPs could be combined with Representative Concentration Pathways (RCPs) which prescribe atmospheric concentrations of GHGs and temperature increase out of the end of the century (see Table 3.1).

Fig. 6.5 is qualitative in nature but more recent work has attempted to quantify these pathways using integrated assessment models (IAMs), which include projections of a wide range of variables, including population growth, economic growth, urbanization, GDP, energy use, land use and GHG emissions, and air pollution for the five SSPs out to the year 2100. A special issue of the journal Global Environmental Change was devoted to this type of quantitative modeling of SSPs (van Vuuren et al., 2017a) and an overview of the results of this modeling is provided by Riahi et al., (2017). Van Vuuren et al. (2017b) compared SSPs 1, 2, and 3, with each other and some of these results are summarized below. The reader may wish to review some of the basic characteristics (or storylines as they are sometimes called) of the five SSPs in Chapter 6 before proceeding. Recall that the standard SSPs invoke no new climate policies, rather they serve as a baseline with which to compare models that do involve new mitigation strategies (e.g., such as a carbon tax). In quantifying the various SSPs, several different IAMs are used to model each SSP – the most representative of these models that best fits the narrative for that SSP is termed the *marker scenario* and many authors tend to focus on that model. The range of the models consistent with each SSP, however, helps in understanding the wide range of uncertainty involved in each SSP. Some results for various SSP-RCP combinations are summarized below:

7.8 Comparison of SSP1 and SSP3

The SSP1 trajectory is related to sustainable development and is characterized by low challenges to both adaptation and mitigation. Part of the narrative is rapid technological development favoring low carbon energy sources (renewables), leading to low emissions making adaptation and mitigation relatively easy. Investment in education and human development are assumed to lead to low population growth and hence low pressure on the land in terms of food demand. Assumptions used in the implementation of the IAMs model include (van Vuuren et al., 2017b): full access to modern energy by 2030; global air pollution will be reduced producing health benefits; significant gains in access to food will also be made. Crop yields, irrigation efficiency, and livestock production are also improved. In this model, the population peaks at about 8 billion around 2050 and then declines to about 6 billion by the end of the century and the GDP growth is strong throughout this period. In the reference scenario for SSP1 (i.e. without new climate policy), the global radiative forcing is about 5 W/m^2 resulting in heating of about 3°C above the preindustrial value by the end of the century. Assuming that a global carbon tax can be agreed upon early on (e.g. by 2020), the SSP1-4.6 and SSP1-2.6 scenarios can be implemented with a relatively small carbon price (<$200/tCO_2$), resulting in heating of between 2.5°C and 1.5°C, respectively. At the time of writing (2021), however, there is little sign of a global agreement on a carbon tax.

In the SSP3 scenario, referred to as Regional Rivalry, emissions are high due to moderate economic growth and rapid population growth and slow technological growth makes adaptation and mitigation in this world challenging. Stress on land use is high leading to less natural landscape. Regional rivalry leads to poor trade flow. The population reaches about 12 billion by the end of the century (twice that of SSP1) and GDP growth is only about a third compared to either SSP1 or SSP2. Global radiative forcing is about 6W/m^2, corresponding to a temperature increase by end of century of about 4.5°C (also similar to the SSP2 scenario). Although the SSP2 storyline is intermediate between SSP1 and SSP3 (Fig. 6.5), most of the metrics outlined above indicate it is closer to SSP3 than SSP1.

7.9 SSP5. Fossil fuel development

In this trajectory, challenges to mitigation are high due to very high emissions but adaptation (Chapter 6) is assumed to be easier on account of rapid technological development. Kriegler et al. (2017) explored the coupled energy, land-use, and emissions scenarios associated with SSP5 which leads to very high fossil fuel use, high energy demand, and CO_2 emissions in the baseline or reference scenario (i.e., no climate mitigation policies). Major land-use changes are implied as demand for crops and livestock increases and forested land area is reduced. The reference scenario is close to RCP8.5 (SSP5-8.5), but the authors above also explore strong mitigation scenarios corresponding to SSP5-4.5 and SSP5-2.6. In the reference scenario, population peaks at about 8.6 billion around 2050 and decreases to 7.4 billion in 2100, not very different from SSP1.

Global energy demand in SSP5 is more than twice than that of SSP1, and end of century economic output (GDP), mainly in the developed countries, is also twice that of SSP1. The emphasis in SSP1 was, however, a sustainable lifestyle, not necessarily rapid growth and to avoid a resource and energy intensive society. Assuming a plentiful supply, fossil fuels dominate the primary energy supply in the baseline SSP5 scenario. In 2050, oil production peaks at twice the level in 2010, and natural gas extraction quadruples due to demand for electricity. Due to rising costs of oil for transportation, coal makes a comeback for electricity generation in the second half of the century. Air pollution levels nevertheless remain low due to effective emissions controls.

In the SSP5 mitigation scenarios (SSP5-4.5 and SSP5-2.6), coal-fired generation is phased out and gas-fired power plants with carbon capture and storage are favored for economic reasons. Biofuels combined with CCS (BECCS) and renewable energy (solar and wind) continue to grow and provide two thirds of energy by 2100. Due to high cost, nuclear power is not seen as a major factor in these mitigation scenarios. In the stringent SSP5-2.6 case, biofuels are greatly expanded to substitute for fossil fuels in combination with CCS. This latter scenario seems highly implausible for a society that until after mid-century still embraced fossil fuels. Some form of catastrophic climate tipping point, however, could force such a radical change in behavior (Kopp et al., 2017). The baseline SSP5 scenario suggests a temperature increase of 5°C over the preindustrial level by 2100 (Kriegler et al., 2017). Such a temperature increase would likely make much of the planet close to uninhabitable (Kopp et al., 2017), and might force a switch to a SSP5-2.6 scenario. These authors point out that the climate impacts and adaptation to SSP5 still remain to be explored. Bill Gates, of Microsoft fame, presented his views of how to avoid a climate disaster in a readable book for a general audience (Gates, 2021). His solutions rely heavily on technological innovation, which are undoubtedly needed. But time is running out.

References

IEA, 2020. CO$_2$ emissions from fuel combustion. https://www.iea.org/commentaries/iea-releases-new-edition-of-global-historical-data-series-for-all-fuels-all-sectors-and-energy-.

Bruckner, T., Bashmakov, I.A., Mulugetta, Y., Chum, H., de la Vega Navarro, A., Edmonds, J., et al., 2014. In: Edenhofer, O., Pichs-Madruga, R., Sokona, Y., Farahani, E., Kadner, S., Seyboth, K., Adler, A., Baum, I., Brunner, S., Eickemeier, P., Kriemann, B., Savolainen, J.,

Schlömer, S., von Stechow, C., Zwickel, T., Minx, J.C. (Eds.), Contribution of Working Group III to the Fifth Assessment Report of the Intergovernmental Panel on Climate Change. Cambridge University Press, Cambridge, UK.

Clarke, l., Jiang, K., Akimoto, K., Babiker, M., Blanford, G., Fisher-Vanden, K., et al., 2014. In: Edenhofer, O., Pichs-Madruga, R., Sokona, Y., Farahani, E., Kadner, S., Seyboth, K., Adler, A., Baum, I., Brunner, S., Eickemeier, P., Kriemann, B., Savolainen, J., Schlömer, S., von Stechow, C., Zwickel, T., Minx, J.C. (Eds.), Contribution of working Group III to the Fifth Assessment Report of the Intergovernmental Panel on Climate Change. Cambridge University Press, Cambridge, UK.

Collins, M., Knutti, R., Arblaster, J., Dufresne, J-L., Fichefet, T., Friedlingstein, P., Gao, X., Gutowski, W.J., Johns, T., Krinner, G., Shongwe, M., Tebaldi, C., Weaver, A.J., Wehner, M., 2013. In: Stocker, T.F., Qin, D., Plattner, G-K., Tignor, M., Allen, S.K., Boschung, J., Nauels, A., Xia, Y., Bex, V., Midgley, P.M. (Eds.), Contribution of Working Group I to the Fifth Assessment Report of the Intergovernmental Panel on Climate Change. Cambridge University Press, UK.

DeAngelo, B., Edmonds, J., J., F., D., W., Sanderson, B.M., 2017. Perspectives on climate change mitigation. In: Wuebbles, D.J., Fahey, D.W., Hibbard, K.A., Dokken, D.J., Stewart, B.C., Maycock, T.K. (Eds.), Climate Science Special Report: Fourth National Climate Change Assessment. U.S. Global Change Research Program, Washington, DC, USA, pp. 584–607.

Edenhofer, O., Pichs-Madruga, R., Sokona, Y., Kadner, S., Minx, J.C., Brunner, S., et al., 2014. In: Edenhofer, O., Pichs-Madruga, R., Sokona, Y., Farahani, E., Kadner, S., Seyboth, K., Adler, A., Baum, I., Brunner, S., Eickemeier, P., Kriemann, B., Savolainen, J., Schlömer, S., von Stechow, C., Zwickel, T., Minx, J.C. (Eds.). Cambridge University Press, Cambridge, UK.

Ehrlich, P.R., Holdren, J.P., 1971. Impact of population growth. Science 171, 1212–1217.

Fischedick, M., Roy, J., Abdel-Aziz, A., Acquaye, A., Allwood, J.M., Ceron, J-P., Geng, Y., Kheshgi, H., et al., 2014. In: Edenhofer, O., Pichs-Madruga, R., Sokona, Y., Farahani, E., Kadner, S., Seyboth, K., Adler, A., Baum, I., Brunner, S., Eickemeier, P., Kriemann, B., Savolainen, J., Schlömer, S., von Stechow, C., Zwickel, T., Minx, J.C. (Eds.), Contribution of Working Group III to the Fifth Assessment Report of the Intergovernmental Panel on Climate Change. Cambridge University Press, UK.

Gates, B., 2021. How to Avoid a Climate Disaster. Knopf, New York and Toronto.

GCCSI, 2020. Global Carbon Capture Institute: Global Status Report. https://www.globalccsinstitute.com/resources/global-status-report. (Accessed December 2020).

IEA, 2020. International Energy Agency. Sustainable Development Scenario. https://www.iea.org/reports/world-energy-model/sustaniable-developments . (Accessed July 2020).

IEA, 2021. Newsletter of the International Energy Agency. The Energy Mix. https://www.iea.org/reports/net-zero-by-2050. (Accessed May 2021).

Kopp, R.E., Hayhoe, K., Easterling, D.R., Hall, T., Horton, R., Kunkel, K.E., 2017. Potential surprises – compound extremes and tipping elements. In: Wuebbles, D.J., Fahey, D.W., Hibbard, K.A., Dokken, D.J., Stewart, B.C., Maycock, T.K. (Eds.), Climate Science Special Report: Fourth National Climate Assessment. U.S. Global Change Research Program, Washington, DC, USA, pp. 608–635.

Kriegler, E., Bauer, N., Popp, A., Humpenöder, F., Leimbach, M., Strefler, J., Baumstark, L., et al., 2017. Fossil-fuelled development (SSP5): an energy and resource intensive scenario for the 21st century. Global Environ. Change 42, 297–315.

LeQuere, C., Jackson, R.B., Jones, M.W., Smith, A.J.P., Abernethy, S., Andrew, R.M., et al., 2020. Temporary reduction in daily global CO_2 emissions during COVID-19 forced confinement. Nat. Clim. Change 10, 647–653.

Lucon, O., Ürge-Vorsatz, D., Ahmed, A.Z., Akbari, H., Bertoldi, P., Cabeza, L.F., Eyre, N., Gadgil, A., et al., 2014. IPCC Working Group III Contribution to AR5. Cambridge University Press, UK.

Myhre, G., Shindell, D., Bréon, F.-M., Collins, W., Fuglestvedt, J., Huang, J., Koch, D., Lamarque, J.-F., Lee, D., Mendoza, B., Naka BPjima, T., Robock, A., Stephens, G., Takemura, T., Zhang, H., 2013. In: Stocker, T.F., Qin, D., Plattner, G.K., Tignor, M., Allen, S.K., Boschung, J., Nauels, A., Xia, Y., Bex, V., Midgley, P.M. (Eds.), Contribution of Working Group 1 to the Fifth Assessment of the Intergovernmental Panel on Climate Change. Cambridge University Press, UK.

Raupach, M.R., Marland, G., Ciais, P., Le Quéré, C., Canadell, J.G., Klepper, G., Field, C.B., 2007. Global and regional drivers of accelerating CO_2 emissions. Proc. Nat. Acad. Sci. 104, 10288–10293.

Rhein, M., Rintoul, S.R., Aoki, S., Campos, E., Chambers, D., Feely, R.A., Gulev, S., Johnson, G.C., Josey, S.A., Kostianoy, A., Mauritzen, C., Roemmich, D. and Talley, L.D. (2013) Observations: Ocean. In, Stocker, T.F., Qin, D., Plattner, G.-K., Tignor, M., Allen, S.K., Boschung, J., Nauels, A., Xia, Y., Bex, V. and Midgley, P.M. (eds.), 2013. *Climate Change: The Physical Science Basis.*Contribution of Working Group I to the Fifth Assessment Report of the Intergovernmental Panel on Climate Change.Cambridge University Press, UK.

Riahi, K., van Vuuren, D.P., Kriegler, E., Edmonds, J., O'Neill, B.C., Fujimori, S., et al., 2017. The shared socioeconomic pathways and their energy, land use and greenhouse gas emissions implications: an overview. Global Environ. Change 42, 153–168.

Sanderson, B.M., O'Neill, B.C., Tebaldi, C., 2016. What would it take to achieve the Paris temperature targets? Geophys. Res. Lett. 43, 7133–7142.

Shine, K.P., Derwent, R.G., Wuebbles, D.J., Morcrette, J.-J., et al., 1990. Radiative forcing of climate. In: Houghton, J.T., Jenkins, G.J., Ephraums, J.J. (Eds.), Climate Change: The Intergovernmental Panel on Climate Change Working Group 1. Cambridge University Press, UK.

Sims, R., Schaeffer, R., Creutzig, F., Cruz-Núñez, X., D'Agosto, M., Dimitriu, D., et al., 2014. In: Edenhofer, O., Pichs-Madruga, R., Sokona, Y., Farahani, E., Kadner, S., Seyboth, K., Adler, A., Baum, I., Brunner, S., Eickemeier, P., Kriemann, B., Savolainen, J., Schlömer, S., von Stechow, C., Zwickel, T., Minx, J.C. (Eds.), Contribution of Working Group III to the Fifth Assessment Report of the Intergovernmental Panel on Climate Change. Cambridge University Press, UK.

UNFCCC, 2015. Paris Agreement. United Nations Framework Convention on Climate Change. https://unfccc.int/files/essential_backfround/convention/application/pdf. (Accessed March 2020).

United Nations, 2020. Sustainable development goals. https://unfoundation.org/what-we-do/issues/sustainable-development. (Accessed July 20200).

Van Vuuren, D.P., Riahi, K., Calvin, K., Dellink, R., Emmerling, J., Fujimori, S., KC, S., Kriegler, E., O'Neill, B., 2017a. The shared socioeconomic pathways: trajectories for human development and global environmental change. Global Environ. Change 42, 148–152.

Van Vuuren, D.P., Stehfest, E., Gernaat, D., Doelman, J.C., van den Berg, M., Harmsen, M., et al., 2017b. Energy, land-use and greenhouse gas emissions trajectories under a green growth paradigm. Global Environ. Change 42, 237–250.

Weinstein, L., Adam, A., 2008. Guesstimation: Solving the World's Problems on the Back of a Cocktail Napkin. Princeton University Press, New Jersey.

PART III

8.   1.5°C versus 2.0°C warming   145
9.   Getting to net zero by 2050   157
10. Climate engineering   167

CHAPTER 8

1.5°C versus 2.0°C warming

8.1 Introduction

As outlined earlier (Chapter 7) the Paris Agreements were signed in 2015 where one of the objectives was to limit the preindustrial level of global warming to 2°C (3.6°F) or less. After the agreement was signed, several countries invited the IPCC to address the impacts to a warming of just 1.5°C, since little information on this amount of warming was available. Recall (Chapter 1) that since preindustrial times the global warming in 2017 was estimated to be 1°C (± 0.2°C), so that a further small warming of only 0.5°C (0.9°F) would reach the 1.5°C benchmark. The IPCC agreed to study this amount of warming and compare it to a warming of 2°C, the result being a Special Report (IPCC, 2018). This chapter is a partial summary of that report.

Recognizing that potential climate change impacts and mitigation actions fall disproportionately on the poor, additional goals of the report were that the emission pathways involve eradication of world poverty and also that they be sustainable – lofty goals to be sure. Although a difference of 0.5°C seems small, the conclusion of the special report showed surprisingly large differences in impacts on human and natural systems and also substantial differences between emission pathway scenarios that limit warming to 1.5°C versus 2.0°C.

Since cumulative GHG emissions vary linearly with temperature the total cumulative emissions since preindustrial times up until some fixed amount of warming can be calculated, then the remaining cumulative emissions budget, necessary for a given temperature rise at a given time, can be estimated. For example, it is estimated that the total cumulative amount of carbon dioxide that could be emitted to keep warming to 2°C is approximately 1000 GtC (DeAngelo et al., 2017). Taking into account the non-CO_2 GHGs this number is reduced to 790 GtC. This means that the remaining cumulative budget is approximately 210 GtC. Assuming the future emissions pathway follows a lower pathway (e.g., RCP4.5; Table 3.1) this threshold is met in 2037, only 16 years away at the time of writing (2021). The constraints to limit warming to 1.5°C are more severe leaving

Climate Change in the Anthropocene.
DOI: https://doi.org/10.1016/B978-0-12-820308-8.00009-X

only 30 GtC for the remaining budget. This threshold would be met in a few years, making it unlikely the world will meet the 1.5°C goal. However, using assumptions somewhat different from above (i.e., warming of 0.2°C per decade since the year 2017) the IPCC Special Report (2018) indicates 1.5°C will be reached in 2040. The latter approach requires fewer assumptions and maybe more reliable. It should be noted that the continents are heating faster than the oceans and heating on land is variable from place to place, such that a substantial faction of the world population is already experiencing impacts from 1.5°C warming (Allen et al., 2018).

A recurring theme in the IPCC (2018) report is overshoot, in which the temperature overshoots the temperature maximum goal by some amount and the temperature subsequently returns to the original goal or lower over the long-term. Overshoot may have irreversible consequences on both human and natural systems (e.g. loss of ecosystems) and it is therefore important to evaluate its magnitude. Keeping warming to 1.5°C will require global transformations under a range of assumptions in areas of: energy, economic growth, technology developments, and lifestyle. To reach a 1.5°C goal, transformations must be more pronounced and occur more rapidly than the 2°C scenario. The current pledges by nations in the Paris accords (referred to as Nationally Determined Contributions (NDCs): range from 52 to 58 $GtCO_{2\ equiv}$/yr. by the year 2030. Emission pathways to 1.5°C, with no overshoot, are much lower, however, and are in the range of 25 to 30 $GtCO_{2equiv}$/yr. (Recall that CO_{2equiv} includes the effects of non-CO_2 GHGs). Pathways that aim for 1.5°C by 2100, with some overshoot, would require extensive large-scale carbon dioxide removal (e.g. afforestation and/or BECCS; see Chapter 10). In that case, anthropogenic CO_2 emissions must decline by 45% from 2010 levels by 2030. In order to limit warming to 2°C, net zero emissions must be reached by 2070, whereas in the case of 1.5°C, net zero should be reached by 2050. This would likely involve a rapid decline in the carbon intensity of electricity generation involving renewable energy.

The difference in risks and impacts of 2°C warming versus 1.5°C on human systems (e.g. health, agriculture) and natural systems (e.g., organisms, ecosystems) is surprisingly large. For example, species projected to lose half of their geographic range under a 2°C warming are: 18% of insects; 16% of plants; 8% of vertebrates. In the case of 1.5°C warming these values reduce to: 6% for insects, 8% for plants, and 4% for vertebrates (Hoegh-Guldberg et al., 2018). Other consequences are extreme weather events, forest fires, invasive species, and pests which all increase at higher warming. Above

1.5°C deserts are also projected to expand and wetlands to be reduced in area. Other benefits to the lower temperature rise are ocean ecosystems, water resources, food security, human health, and fewer extreme climate events.

Restricting warming to 1.5°C involves a global response (involving major societal transformations) that is characterized by several asymmetries. One is the differential contribution to global warming. Developed nations have contributed more to the problem and bear more responsibility compared to developing nations. Conversely, developing nations are likely to be more adversely affected than developed nations. Poor nations are less able to adapt to climate change thereby acting as a threat multiplier and compounding other drivers of poverty. Developing nations may find it more difficult to reduce emissions at a time when they are seeking greater development, resulting in an inherent conflict. Limiting warming to 1.5°C is therefore a very ambitious project (and not even plausible given actions to date).

8.2 1.5°C and 2.0°C warming

The 1.5°C warming in the IPCC special report (Allen et al., 2018) refers to the global mean surface temperature (GMST) at a given time above preindustrial levels including both sea and land centered over a thirty year period. The method of calculation of GMST was outlined in Chapter 1 and it excludes natural fluctuations in climate, due to volcanoes, for example. The preindustrial reference is taken as the mean over the period 1850 to1900 in order to be consistent with IPCC-AR5 (2013). As mentioned above, using these criteria, the warming as of 2017 is estimated to be 1.0°C (±0.2°C). Fig. 8.1 shows a variety of pathways to 1.5°C warming over the period 2010-2100 based on cumulative emissions. The uppermost curve shows overshoot that eventually stabilizes at 1.5°C.

Because of a lack of data the IPCC Special Report (2018) focuses on the transient response to warming rather than the long-term equilibrium response which would take millennia to be realized. In addition, artificial modification of the climate (see Chapter 10) is not addressed in the report. Again due to lack of data, the interaction of other drivers of poverty together with climate change is not addressed in the report.

A simple way to compare risks due to 1.5°C warming versus 2.0°C warming is to look at the historically observed warming response of 0.5°C over the past 50 years and assume a linear response to further warming

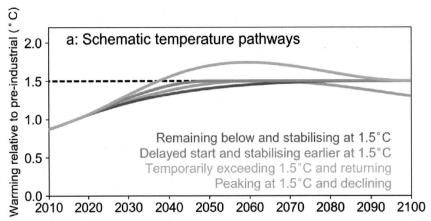

Figure 8.1 Relationship of different 1.5°C temperature pathways reflecting cumulative emissions overtime. Overshoot with a return to 1.5°C (blue line, upper curve) is shown. The lower curve (brown) peaks at 1.5°C and then declines. *(Source with permission: Fig. 1.4, Allen et al., 2018.)*

(which may not be appropriate). As mentioned already, there does appear, however, to be a linear temperature response to cumulative CO_2 emissions. In addition to observed responses, climate models using different RCP pathways (Table 3.1) are commonly used. When observed responses agree with climate models, some confidence (classified as medium, high or low) can be placed in the results.

In addressing transient responses, three factors need to be considered regarding overshoot: the maximum temperature reached, the length of the overshoot period, and the rate of change during the overshoot. Risks to human and natural systems are then estimated using impact models, driven by input from climate models. Recall that risk refers to the probability of occurrence of hazardous events multiplied by the impacts if such events occurred (Chapter 7). Impact models depend on temporal aspects (e.g., overshoot or not) and on RCP pathways. It is a challenge to downscale from global results to smaller human scale systems or ecosystem level systems because of a complex combination of local and global drivers. In addition, it is difficult to associate uncertainties to each step in the methodology.

Fig. 8.2 (top panel) shows the global mean temperature for a warming of 1.5°C and 2.0°C warming and the difference between the two (2.0°C–1.5°C). The difference diagram shows warming on all of the continents, especially at high latitudes relative to 1.5°C. The bottom panel in Fig. 8.2 shows the same results for precipitation. Most locations show higher precipitation

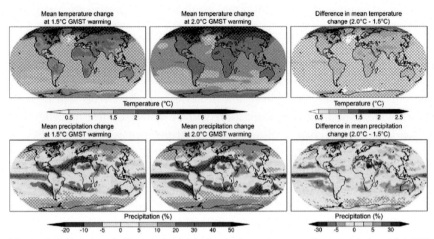

Figure 8.2 Top panel. Shows the global mean temperature for a warming of 1.5°C and 2.0°C warming and the difference between the two (2.0°C–1.5°C). The difference diagram shows warming on all of the continents, especially at high latitudes relative to 1.5°C. Bottom panel. Shows the same results for precipitation. Most locations show higher precipitation at higher warming, as expected, but the results are not as robust as those for temperature. *(Source with permission: Fig. 3.3, Hoegh-Guldberg et al., 2018.)*

at higher warming, as expected, but the results are not as robust as those for temperature. Some risks to natural and human systems are presented below (Hoegh-Guldberg et al., 2018). Climate models are better at projecting climate change for a given amount of radiative forcing rather than projecting risk for differential temperatures (such as 1.5°C versus 2.0°C).

8.3 Natural systems

Sea level. Sea level can be addressed on different temporal and spatial scales, including global mean sea level (GMSL), regional variations about this mean, and sea level extremes due to storm surges and tides. Sea level has been rising since the late nineteenth century. Reports of slowing sea level rise are attributed to satellite instrument drift. Correcting for this, GMSL rise is estimated to be 2.6 to 2.9 mm/yr and this rise includes contributions from thermal expansion and ice loss on land and freshwater storage. Following a low emissions pathway (RCP2.6; Table 3.1) models indicate little difference in the contributions of glaciers between 1.5°C and 2.0°C worlds, largely due to the slow response time of glaciers to warming. By the end of the century a GMSL rise of 0.3 to 0.6 m is projected for RCP2.6. Under this pathway it is estimated that one-in-a-hundred year coastal flood would quadruple compared to no sea level rise. It is also estimated that sea level rise would be 0.1m less by the end of the century compared to the year 2000 for 1.5°C compared to 2.0°C.

Sea level rise is already causing salinization, flooding, and erosion and is projected to affect human and ecological systems, including freshwater systems, biodiversity, agriculture, and fisheries and the effects will be seen worldwide. Flooding and salinization will be especially felt in deltas and estuaries. Climate change mitigation and adaptations would likely reduce or delay coastal impacts. Without any action the cost of these impacts is likely to be in the thousands of billions (USD). Due to the commitment to SLR, because of past emissions, the uncertainties in projections for 1.5°C and 2.0°C warming overlap with each other. Assuming no adaptation or additional protection, by 2100, 31–69 million people will be exposed to flooding for 1.5°C warming and 32–79 million people for a 2.0°C warming. Risks are projected to be highest for south and Southeast Asia. With the exception of New York, Amsterdam, and London there is limited data on how cities will cope with higher sea level on a century time scale.

Freshwater. Global trends in stream flow since the 1950s are not statistically significant. In the case of water availability and drought most models shows increased drought and less water availability for southern Europe and the Mediterranean region (including the Middle East) and southern Africa with a shift from 1.5°C to 2.0°C. These regions already show drying trends. In the Middle East the drought of 2008 was a high temperature, low precipitation event, whereas the drought of 1960 was a low temperature, low precipitation event. The population saw rapid growth between these periods and a decline in agriculture production due to drought, with consequent migration to cities. A similar pattern maybe expected with a shift from 1.5°C to 2.0°C.

Biome shifts. Biome shifts (major ecosystem types) in latitude and elevation (e.g. shrub encroachment on tundra) have been detected in boreal, temperate, and tropical regions and these are attributed to climate change more so than other factors. An ensemble of general circulation models (GCMs) that include dynamic vegetation variation indicate that 13% of biomes are transformed at 2.0°C warming but only 6.5% at 1.5°C% warming. If warming is limited to 1.5°C, biome shifts in Tibet, Himalyas, Arctic, southern Africa, and Australia would be avoided.

Phenology and species range. Phenology is the cyclic or seasonal variation in plant and animal life and it is a sensitive indicator of climate change. Examples include flowering times of plants, migration of birds, egg laying, and emergence of butterflies. Advancement of about 3 days has been observed in the Spring phenology of plants and animals in the northern hemisphere (between 30 and 70°N). The Arctic is strongly affected because of the faster warming there, whereas the Tropics appear to be more affected by moisture stress. Ecosystem disruption may occur by species-species interaction that has different response times, such as insect plant pollinators (referred to as phonological mismatch). The spring advancement is estimated to be about 10 days in the case of 2.0°C warming compared to 1.5°C. Some plants, however, may be able to mitigate the effects of warming by expanding their range.

The percentage of species that would lose half of their range was pointed out in the introduction to this chapter and the numbers, based on the study of 105,500 species, bears repeating: species projected to lose half of their geographic range under a 2°C warming are: 18% of insects, 16% of plants, 8% of vertebrates. In the case of 1.5°C warming these values reduce to 6% for insects, 8% for plants, and 4% for vertebrates.

Oceans. The oceans play an important role in regulating atmospheric gas concentrations, global temperatures and climate. It also provides habitat to a large number of organisms and ecosystems that provide goods and services of enormous economic value. Global warming poses a threat to a large number of ocean ecosystems (e.g., tropical coral reefs) with implications for the livelihood, food source, and safety of many coastal communities worldwide. Climate change has resulted in an increase in temperature of the upper 700 m of the ocean, and changes in chemistry (decrease in pH and oxygen content), and ocean stratification which affects circulation, mixing and salinity. Warmer water has caused volumetric expansion and sea level rise and increased intensity of storms in some regions and reduction in sea ice volume. These changes will continue at 1.5°C warming and more so at 2.0°C warming above preindustrial levels. Chapter 5 looked at pH change and deoxygenation and also temperature extremes in the ocean.

Ecosystems and food webs. Marine organisms that build physical structures ('ecosystem engineers') include seagrass, kelp, oysters, salt marsh species, mangroves and corals (e.g. sea grass meadows, kelp and mangrove forests, oyster and coral reefs). These organisms provide food, livelihoods, and coastal protection. Many of these organisms are sensitive to sea level, temperature extremes, storms, and sediment supply. The Great Barrier Reef has lost 50 percent of its shallow water corals along hundreds of kilometers of its length. The reef has also undergone several heat-related bleaching events since the 1980s. At a warming level of 1.3°C it is projected that coral reefs are at high risk and a warming of 1.8°C leads to a high level of risk for sea grasses (due to sea level rise, erosion, extreme temperatures and storms). As sea level rises many of these ecosystems, such as salt marshes, mangroves, and seagrasses, are forced inland where they encounter coastal infrastructure. Future coastal zone management should recognize the importance of these ecosystems for protection of coastal communities.

Representative groups of marine organisms that are part of the vast oceanic foodweb are an important source of food for humans and other organisms that include: bivalve mollusks (clams, oysters, and mussels) which are filter feeding invertebrates; pteropods which are small pelagic mollusks that filter feed in suspension and are a food source for fish, whales, and seabirds; finned fish are important to fisheries and an important source of food for humans globally; krill are a food source for seabirds and whales. Ocean acidification (see Chapter 5) and a temperature increase of 1.5°C puts many of these and associated organisms at moderate to high risk. Dissolution of calcium carbonate shells (aragonite) has already been demonstrated in the case of mollusks. Finned fish are expected to show moderate risk at 1.3°C and very high risk at 1.8°C warming. Many species of fish and lobster have already moved northward or into deeper cooler water (Chapter 5).

Sea Ice. Arctic sea ice decreased over the period 1997–2014 at a rate of 130,000 km^2/yr. This is four times faster compared to the period 1979–1996 when satellite observations first began. Ice thickness also decreased by about 50 percent in the central Arctic. Climate models also show ice loss and thinning. However, CMIP5 models (Coupled Model Intercomparison Project phase 5; see Chapter 3) show smaller losses than those observed. This mismatch indicates that model projections are likely to underestimate ice loss for a given amount of warming. Correcting for this bias, for 1.5°C warming, the models project the Arctic will maintain ice cover in the summer. For 2.0°C warming the chances for an ice-free Arctic are substantially higher. After 10 years of stabilized warming at 2.0°C there is a chance of at least one ice-free summer. In the case of 1.5°C, 100 years of stabilized warming would produce one ice-free summer. More recent models indicate the Arctic will be ice-free in summer after 1.0°C warming over present day conditions. During winter, little loss of ice is projected for either 1.5°C or 2.0°C. In the case of temperature overshoot, no hysteresis was observed, so that warming and cooling were indistinguishable. The situation for Antarctic sea ice is regionally more complex with losses and gains depending on location (see Chapter 1).

8.4 Human systems

Agriculture: crop production. Depending on crop type and geographic area, many non-climate variables interact with climate change to affect crop production. Despite these uncertainties, studies indicate climate change has already affected the suitability of crops resulting in changes in production levels in many parts of the world. Temperature and precipitation trends have reduced crop production and yields negatively for cereals (maize and wheat), whereas the effects on rice and soybean yields are less clear, being positive or negative. Warming has resulted in positive effects in some high latitude areas (e.g., Siberia), making it possible to expand the growing season with the possibility of more than one harvest per year in some areas.

Increase in atmospheric carbon dioxide concentration increases yields by enhanced radiation and water use efficiencies (the carbon dioxide fertilization effect), but elevated levels of carbon dioxide also coincide with lower protein contents for cereals and also Asian varieties of rice. This could result in undernourishment for a large segment of the global population. Overall, elevated levels of CO_2 is expected to have a negative impact on global food production.

Crop yields in the future are also expected to be affected by changes in temperature and precipitation. At low latitudes, studies indicate wheat and maize have lower yields at a warming of 1.0 to 2.0°C. Constraining warming to 1.5°C rather than 2.0°C would avoid significant risks of declining crop yields in West Africa, Southeast Asia and Central and South America. A significant reduction of about 5 percent has been projected in global production of wheat, rice, maize and soybean for each 1.0°C rise in global mean temperature. Other indirect effects include pests and diseases.

Human health. Increased morbidity and mortality are associated with extreme weather and climate events such as drought, wildfires, heavy rainfall, floods, and storm surges. For example, hurricane Harvey (Texas, 2017) caused the death of 59 individuals. These events can adversely affect the physical and mental health of individuals (Chapter 5). The health care system itself may be disrupted by extreme weather events. Temperature extremes affect different groups in the population, with children, older adults, pregnant women and outdoor workers most susceptible. Heat-related deaths outweigh those due to cold-related deaths for most regions. However, in February of 2021, a freak Arctic vortex escaped southward to Texas where temperatures reached -15°C (5°F). The entire state electric grid failed due to energy demand and

201 people lost their lives from a variety of causes, but mainly hypothermia (NYT, 2021).

Morbidity and mortality are projected to increase for warming of 1.5°C and more so for 2.0°C depending on region, built environment, acclimatization of the population and access to air conditioning. Work place heat-related illness will increase and outdoor work productivity will decline. Because ozone formation is temperature dependant, air quality will decrease and ozone-related mortality will increase. The geographic and seasonal range of disease bearing mosquitos (e.g., dengue fever, malaria) is predicted to increase as will tick-borne diseases such as Lyme disease.

Urban areas. The heat island effect (Chapter 5) in urban areas will increase the frequency of heat-related extremes with greater effects on vulnerable populations. Extreme events in megacities are projected to affect 350 million people by 2050 assuming midrange population growth. As cities grow the heat island effect is projected to also grow. Infrastructure such as water, energy, buildings, and transportation will be stressed. Thermal power stations may go to partial power due to lack of cooling water. Energy demand will increase due to increased air conditioning. Adaptation actions may included making roofs more reflective and incorporating more shade. Studies have shown that poorer neighborhoods in cities have less green space and less shade.

Tourism. Climate change is expected to have far-reaching consequences for tourism markets and destinations. Factors to be considered are: heat waves, hurricanes, wild fires, reduced snow pack, coastal erosion, coral reef bleaching which are all expected to increase in frequency. In Europe, based on analyses of tourist comfort, overnight stays are expected to increase for northern Europe and decrease for southern Europe for a 1.5°C warming. Two degree warming is projected to result in a 5% decrease in European tourism and as high as a 10 percent decrease for southern Europe with potential gain for UK.

In summary there is growing evidence that there are increased risks across a broad range of natural and human systems for warming of 1.5°C and 2.0°C above preindustrial levels. The simplest way to judge these risks has been to evaluate the effects observed for the 0.5°C warming over the past 50 years. The difference in risks involved between these two temperature goals is surprisingly large. The next chapter addresses mitigation and adaptation strategies necessary in order to meet the goal of net zero emissions by 2050 to avoid a 2.0°C warming.

References

Allen, M.R., Dube, O.P., Solecki, W., Aragón-Durand, F., Cramer, W., Humphreys, S., Kainuma, M., Kala, J., Mahowald, N., Mulugetta, Y., Perez, R., Wairiu, M., Zickfeld, K., 2018. Framing and context. In: Masson, D.V., Zhai, P., Pörtner, H.-O., Roberts, D., Skea, J., Shukla, P.R., Pirani, A., Moufouma-Okia, W., Péan, C., Pidcock, R., Connors, S., Matthews, J.B.R., Chen, Y., Zhou, X., Gomis, M.I., Lonnoy, E., Maycock, T., Tignor, M., Waterfield, T. (Eds.), Global Warming of 1.5°C. An IPCC Special Report on the impacts of global warming of 1.5°C above pre-industrial levels and related global greenhouse gas emission pathways, in the context of strengthening the global response to the threat of climate change, sustainable development, and efforts to eradicate poverty. Cambridge Press, UK.

DeAngelo, B., Edmonds, J., Fahey, D.W., Sanderson, B.M., 2017. 2017. Perspective on climate mitigation. In: Wuebbles, D.J., Fahey, D.W., Hibbard, K.A., Dokken, D.J., Stewart, B.C., Maycock, T.K. (Eds.), Climate Science Special Report: A Sustained Assessment Activity of the U.S. Global Change Research Program. U.S. Global Change Research Program, Washington, DC, USA, pp. 584–607.

Hoegh-Guldberg, O., Jacob, D., Taylor, M., Bindi, M., Brown, S., Camilloni, I., Diedhiou, A., Djalante, R., Ebi, K.L., Engelbrecht, F., Guiot, J., Hijioka, Y., Mehrotra, S., Payne, A., Seneviratne, S.I., Thomas, A., Warren, R., Zhou, G., 2018. Impacts of 1.5°C global warming on natural and human systems. In: MassonDelmotte, V., Zhai, P., Pörtner, H.-O., Roberts, D., Skea, J., Shukla, P.R., Pirani, A., Moufouma-Okia, W., Péan, C., Pidcock, R., Connors, S., Matthews, J.B.R., Chen, Y., Zhou, X., Gomis, M.I., Lonnoy, E., Maycock, T., Tignor, M., Waterfield, T. (Eds.), Global Warming of 1.5°C. An IPCC Special Report on the impacts of global warming of 1.5°C above pre-industrial levels and related global greenhouse gas emission pathways, in the context of strengthening the global response to the threat of climate change, sustainable development, and efforts to eradicate poverty. Cambridge Press, UK.

IPCC, 2018. Global warming of 1.5°C. In: MassonDelmotte, V., Zhai, P., Pörtner, H.-O., Roberts, D., Skea, J., Shukla, P.R., Pirani, A., Moufouma-Okia, W., Péan, C., Pidcock, R., Connors, S., Matthews, J.B.R., Chen, Y., Zhou, X., Gomis, M.I., Lonnoy, E., Maycock, T., Tignor, M., Waterfield, T. (Eds.), An IPCC Special Report on the impacts of global warming of 1.5°C above pre-industrial levels and related global greenhouse gas emission pathways, in the context of strengthening the global response to the threat of climate change, sustainable development, and efforts to eradicate poverty. Cambridge Press, UK.

NYT, 2021. Texas Study Shows Toll From Storm was Higher. New York Times, p. A17.

CHAPTER 9

Getting to net zero by 2050

9.1 Introduction

The International Energy Agency (IEA) is an intergovernmental organization that was established in 1974 in response to the oil embargo of 1973. It was initially formed to respond to disruptions of oil supply, for example by releasing stockpiles, and to provide statistical information on the international oil market to the fossil fuel industry, and as a counterbalance to OPEC. It has since broadened its permit to include renewable energy sources with a view to a global sustainable energy sector. Its member countries are mainly consumers, not producers, of oil and gas. In the past, the IEA has been criticized by environmental groups for downplaying the potential of renewable sources of energy, but it appears they may have rectified this bias (Wikipedia, 2021). In 2021, the IEA released a special report titled: Net Zero by 2050: A Roadmap for the Global Energy Sector (IEA, 2021). The goal of the report is to maintain global warming to 1.5°C above preindustrial levels. Parenthetically, it was pointed out in the previous chapter, that assuming a warming of 0.2°C per decade since 2017 (an IPCC assumption), net zero emissions would need to be reached by 2040. Nevertheless, the IEA report (224 pages in length) is ambitious and consists of four chapters:

1. It explores the outlook for global CO_2 emissions based on current policies and pledges.
2. Describes how energy demand needs to evolve to meet net zero emissions by 2050 (NZE).
3. Examines alternative energy sources for various sectors of the economy.
4. Explores NZE for the energy industry, the economy, citizens, and governments.

This chapter is a partial summary of the IEA report (IEA, 2021). Three principles or assumptions underlie the report:

- Technology options and emissions reduction options are dictated by costs, technology maturity, policy preferences, and market and country conditions.
- All countries co-operate toward achieving net-zero worldwide emissions.

Climate Change in the Anthropocene.
DOI: https://doi.org/10.1016/B978-0-12-820308-8.00006-4

- An orderly transition in the energy sector – this includes security of fuel and electricity supplies and avoiding volatility in the energy market.

With the publication of the new IPCC-AR6 (2021) report in August of 2021, it became clear that the second assumption above is largely wishful thinking (NYT, 2021). The more vulnerable nations maintain the rich industrialized nations must immediately reduce their emissions and compensate poor nations for damages already caused and fund preparation for future climate events. Whether advanced economies will heed this appeal is highly uncertain. Within the IEA report the key pillars of decarbonization are as follows: energy efficiency, behavorial changes, electrification, renewables, hydrogen and hydrogen-based (e.g., ammonia) fuels, bioenergy and CCUS. Within the category of behavioral changes, examples include riding public transport, buying electric cars, installing heat pumps in residences.

9.2 The current situation (2021)

There has been a rapid rise in the number of governments worldwide who have pledged to reduce greenhouse gas emissions to net zero. However, few of these pledges are fixed in domestic legislation and few are backed up by specific policies to deliver them on time and in full. One way to understand where the world is at present (2021) relative to net-zero emissions by 2050 is to examine two different scenarios: first, what the IEA roadmap calls STEPS, referring to Stated Policy Scenarios, that have been announced by governments that are currently in place, and second, what is termed APC, or Announced Pledge Cases, which assumes that all pledges will be met in time and in full, whether they are underpinned by policies or not. The APC scenario includes pledges made by nationally determined contributions (NDCs) under the Paris Agreement. As might be expected, the APC scenario goes further than STEPS in meeting net zero goals, but both scenarios fall well short of net-zero by 2050.

In the case of STEPS, the annual energy-related and industrial processes emissions rise from 34Gt in 2020 to 36Gt in 2030 and remain at this level until 2050. If emissions continue on this trajectory this would lead to a rise in temperature of 2.7°C (with a 50 % probability) by the end of the century. Renewables provide 55 percent of global electricity production in 2050 but clean energy lags in other sectors. Global coal use falls by 15 percent between 2020 and 2050 and natural gas use is 50% higher.

In the case of APC, global energy-related and industrial processes fall to 30Gt in 2030 and to 22Gt in 2050. These emissions would lead to a

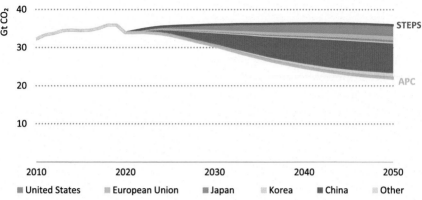

Figure 9.1 Reduction in CO_2 emissions for two different scenarios (STEPS and APC; see text) by region over time. *(Source: Fig. 1.10, IEA, 2021.)*

rise of 2.1°C (50% probability) by the end of the century. The share of renewable rises to 70% in 2050. Coal use drops by 50 percent in 2050, while natural gas expands 10 percent in 2025 and remains level until 2050. Efficiency, electrification, and replacement of coal by low emissions sources play a central role in the APC net zero goals. The relative contribution of nuclear, hydrogen, bioenergy, and carbon capture utilization and storage (CCUS) vary across countries depending on their circumstances. Fig. 9.1 shows the divergence between STEPS and APC over time and across different countries and underscores the importance of net-zero pledges in APC. As mentioned already, both scenarios fall well short of net zero by 2050.

Under the Paris Agreement (UNFCCC, 2015; Chapter 7), nations are required to submit Nationally Determined Contributions (NDCs) representing the amount of their intended reduction in emissions with policies to meet their goals. Some of these pledges came on condition of international financial support or technological support. One hundred and ninety one countries signed the Paris Agreement, representing 90% of energy-related and industrial global emissions. The global pandemic of 2020 caused a decrease in emissions of about 5%, but emissions rapidly rebounded in 2021 to near 2019 levels after the economy open up. As of April 2021, 80 countries submitted updated pledges of emissions, most of which were more stringent than the original NDCs, covering about 40% of global emissions. Of these countries, 44 pledged to meet net-zero emissions by about mid-century. Most of these pledges are not backed up by detailed policies or implementation plans and are not included in STEPS.

It is useful to look at electricity generation in the APC scenario as its emissions are lower than in STEPS and closer to net-zero by 2050. Electricity generation doubles in the next three decades by low-emission sources. By 2050 half of electricity will be by solar PV (photo voltaic) and wind. Hydropower continues to expand and equals the third largest source by 2050. Nuclear power also continues to expand up to 10 percent, led by China. Coal demand falls to less than 10% in 2050. Remaining coal-fired plants are equipped with CCUS. Ammonia and hydrogen emerge as fuels in electricity generation by 2030. Total battery capacity rises substantially, reaching 1600 Gw in 2050.

9.3 Road to net-zero emissions 2050

The following is a summary of NZE in the energy sector. In recent years the energy sector was responsible for about three quarters of global emissions.

- In NZE, global energy-related and industrial processes, CO_2 emissions fall by 40% between 2020 and 2030 and reach net-zero in 2050. Universal access to sustainable energy is achieved by 2030. These changes take place while GDP doubles by 2050 and population increases by 2 billion to about 9.7 billion.
- Total energy supply falls by 7% between 2020 and 2030 and remains at this level into 2050. The leading sources of electricity are solar PV and wind by 2030 and by 2050 represent 70 percent of global generation. The traditional use of bioenergy (e.g., cooking) is phased out by 2030.
- Coal demand declines by 90% in 2050, oil declines by 75%, and natural gas by 555. The remaining fossil fuels are used in industrial processes where low-emission technology options are scarce (e.g., cement).
- In NZE, energy efficiency, and wind and solar account for half of emissions savings to 2030. The period 2030 to 2050 sees increased electrification, hydrogen and deployment of CCUS. The latter two technologies are not yet currently available at scale. Behavior changes by individuals and businesses account for a reduction in CO_2 emissions of 1.7Gt by 2030.
- Investment in the energy sector, which was in recent years $2.3 trillion, climbs to $5 trillion by 2030. By 2050 this investment would represent about 1% of global GDP.

9.4 Population and GDP

That the population will be 2 billion more in 2050 (see above) is in line with the United Nations projections with most of the growth in emerging market and developing economies. From 2022 (post Covid pandemic) the GDP growth is assumed to be 3%, in line with IMF estimates. Governments have large debt due to the pandemic, but interest rates are low so that this is manageable in the long term. By 2030, the economy is 45% larger and by 2050 it is twice as large.

9.5 Energy and CO_2 prices

A broad range of energy policies and accompanying measures are introduced across all regions to reduce emissions in NZE. They include (IEA, 2021): renewable fuel mandates, efficiency standards, market reforms, research, development and deployment, and elimination of fossil fuel subsides. In the transport sector electric vehicles replace the internal combustion engine, and there is increased use of liquid and synthetic fuels in aviation and shipping.

Projections of energy prices are challenging at best. The NZE assumes a supply and demand equilibrium with the goal of avoiding volatility. The rapid drop in oil and gas demand means that no new fields need be discovered and this also applies to coal mines. The lower demand will extend the lifespan of existing resources. The NZE anticipates that oil prices drop to $35 per barrel by 2030 and $25 per barrel by 2050 (for comparison crude oil was $67 per barrel in august 2021). Much depends on what producers such as OPEC decide to do – decrease production to raise prices or increase production to gain more income. The latter option has been common in the past. However, with low demand and plentiful supply, prices could spiral downward, increasing volatility. Alternatively, producers could use their resources to produce low-carbon fuels, such as hydrogen, for export.

Carbon dioxide prices are introduced across all regions with three different scales depending on the stage of economic development, and increasing as time goes on (Table 9.1). By way of comparison, Europe pays 50 Euros/tonne ($60 USD in 2021). For emerging markets it is anticipated that these economies will go directly to renewable sources and so pay less for emissions.

Table 9.1 CO_2 prices for electricity, industry, and energy production.

Price/tonne (USD)	2025	2030	2040	2050
Advanced economies	75	130	205	250
Developing economies (BRICS)	45	90	160	200
Emerging markets	3	15	35	55

BRICS, Brazil, Russia, India, China.

9.6 CO_2 emissions

In 2019 global energy-related CO_2 emissions were 33.4 Gt but decreased to 3.15 Gt during the global pandemic of 2020. Global energy-related and industrial emissions in the NZE report decrease to 21 Gt in 2030 and to net-zero in 2050. Developing nations are anticipated to have large renewable resources that can be used for CCUS (e.g. biofuels) and direct air capture (DACCS; Chapter 7). Per capita emissions in advanced economies drop from about 8 tCO_2 per person to 3.3 tCO_2, similar to that in developing nations in 2020. By 2040 per capita emissions in both developing and developed regions is projected to be 0.5 tCO_2 per person. Cumulative global energy-related and industrial emissions are estimated at 500Gt, including forestry, agriculture and land changes, between 2020 and 2050. This estimate is in line with IPCC Special Report (2018) to keep warming to 1.5°C with 50 percent probability. Methane emissions fall from 115 million tonnes (Mt) in 2020 to30 Mt in 2030 and 10Mt in 2050. Due to the short residence time of methane in the atmosphere ($\sim$ a decade) these reductions will have significant effects on warming in the short term. One tonne of methane is equivalent to 30 tonnes of CO_2.

9.7 Total energy supply

Total energy supply falls to 550 exajoules (1 EJ $= 10^{18}$ J) in 2030, 7% lower than in 2020. To give the reader an idea of what an exajoule is: the total global earthquake energy expended annually is between 10^{18} and 10^{19} J, and the total solar energy intercepted by the Earth annually is 5 x 10^{24} J or five million EJ (Brown, 1981). In 2020 the energy supply was dominated by coal, oil, gas, and nuclear (Fig. 9.2). Despite the increase in population and the economy the decrease in 2030 is attributed to a fall in energy intensity (cost of energy). This is achieved by electrification, energy efficiencies, behavorial changes that reduce demand for energy services, and a shift away from traditional use of bioenergy (e.g. switch from cooking using biomass to more

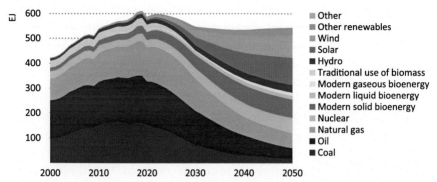

Figure 9.2 Total energy supply for net-zero emissions by 2050. Renewables and nuclear energy largely replace fossil fuels by 2050. *(Source: Fig. 2.5, IEA, 2021.)*

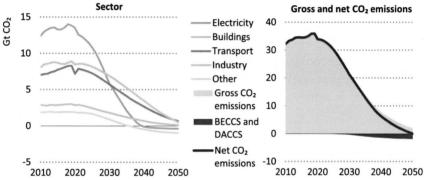

Figure 9.3 CO_2 emissions by sector. Emissions fall fastest in the power sector. *(Source: Fig. 3.1, IEA, 2021.)*

efficient natural gas). The decline in energy efficiency between 2030 and 2050 declines to 2.7 percent per year. Total energy supply in 2050 is similar to that of 2010, despite a 3 billion greater population and an economy three times larger. In 2050, two-thirds of energy is supplied by renewables such as bioenergy, wind, solar, hydroelectric and geothermal.

9.8 Economic sectors

The fastest and largest reductions in global emissions in the NZE are initially in the electricity sector (Fig. 9.3.) In 2020 electricity generation was the largest sources of emissions and drop by nearly 60% by 2030 mainly due to reduction in coal-fired plants (replaced by low-emission biomass plants). By 2050, nearly 90% of electricity generation is from renewables.

In the industrial sector emissions reductions during the period 2030 are through energy and materials efficiency improvement, electrification of heat and fuel switching to solar thermal, geothermal, and bioenergy. After 2030 CCUS and hydrogen play an increasingly important role, especially in heavy industries, such as cement, steel, and chemicals.

In road transport electricity dominates the sector making up 60% of energy in 2050. Hydrogen-based fuels play a role in long-haul heavy-duty trucks. Hydrogen-based fuels, such as ammonia, increasingly displace oil. In aviation and shipping synthetic liquids and advanced biofuels grow rapidly.

In buildings, electrification of heating accounts for two-thirds of total building sector energy consumption. By 2050, most residential buildings in advanced economies and 40% in emerging markets economies are fitted with heat pumps. Solar water heaters and biomass boilers provide a further quarter of final energy use in the building sector in 2050. Building energy falls by 25% between 2020 and 2030 largely due to improved efficiency (improved building envelopes and retrofitting existing buildings) and phase out of traditional cooking with solid biomass, replaced with liquid petroleum gas, biogas, electric cookers, and efficient bioenergy stoves. Universal access to electricity is achieved by 2030, adding only 1% total to global energy demand.

9.9 Conclusions

The apparent lack of international cooperation, as mentioned earlier, is a major problem in implementing either the IEA (2021) scenario or the IPCC (2018) scenarios, such that reaching net zero by 2050 appears, to this author, highly unlikely. An additional problem is that while many governments propose net-zero emissions by mid-century, these governments are still issuing oil and gas leases for additional exploration. This amounts to a contradiction since burning additional fossil fuels and net-zero emissions are incompatible. Given that net-zero emissions by 2050 is unlikely, a last resort alternative is climate engineering, which most scientists do not support. The final chapter examines climate engineering or geoengineering.

References

Brown, G.C., 1981. The Energy Budget of the Earth. The Cambridge Encyclopedia of Earth Sciences. Cambridge Press, Cambridge, UK.
IEA, 2021. Net Zero by 2050. IEA, Paris. https://www.iea.org/reports/net-zero-by-2050. All rights reserved. Accessed August 2021.

IPCC, 2018. In: Masson, Delmotte V., Zhai, P., Pörtner, H.-O., Roberts, D., Skea, J., Shukla, P.R., Pirani, A., Moufouma-Okia, W., Péan, C., Pidcock, R., Connors, S., Matthews, J.B.R., Chen, Y., Zhou, X., Gomis, M.I., Lonnoy, E., Maycock, T., Tignor, M., Waterfield, T. (Eds.), Global Warming of 1.5°C. An IPCC Special Report on the impacts of global warming of 1.5°C above pre-industrial levels and related global greenhouse gas emission pathways, in the context of strengthening the global response to the threat of climate change, sustainable development, and efforts to eradicate poverty. Cambridge Press, UK.

IPCC-AR6, 2021. In: Masson, Delmotte, V., Zhai, P., Pirani, A., Connors, S.L., Péan, C., Berger, S., Caud, N., Chen, Y., Goldfarb, L., Gomis, M.I., Huang, M., Leitzell, K., Lonnoy, E., Matthews, J.B.R., Maycock, T.K., Waterfield, T., Yelekçi, O., Yu, R., Zhou, B. (Eds.), Climate Change 2021: The Physical Science Basis. Contribution of Working Group I to the Sixth Assessment Report of the Intergovernmental Panel on Climate Change. Cambridge University Press, Cambridge, USA.

NYT, 2021. Have-Nots Tell Richer Nations to Fix Climate. New York Times, p. A1.

UNFCCC, 2015. Paris Agreement. United Nations Framework Convention on Climate Change. https://unfccc.int/files/essential_background/convention/application/pdf. (Accessed March 2020).

Wikipedia, 2021. https://en.wikipedia.org/wiki/International_Energy_Agency. Accessed Jan. 2021.

CHAPTER 10

Climate engineering

10.1 Introduction

Climate engineering (also called geoengineering) refers to the intentional modification of the Earth's climate and it has been proposed in order to mitigate the climate response to elevated greenhouse gas (GHG) emissions. The term climate engineering was first used by Marchetti (1977) where it was proposed that carbon dioxide could be disposed of in the deep ocean. Given the lack of progress in international agreements in mitigating GHG emissions and the increasing frequency of extreme climate events there has been renewed interest in climate engineering research (NYT, 2021). Climate engineering is commonly divided into two broad categories: solar radiation management (SRM) in which a fraction of solar insolation to the Earth is redirected back to space, commonly by artificially increasing the Earth's albedo, thereby counteracting the heating effects of GHGs. During deployment of SRM, however, GHGs continue to rise. It is thought that SRM in general would be a relatively low cost option (Barrett, 2008), but high risk (Matthews and Caldeira, 2007; Hegerl and Solomon, 2009), in terms of environmental consequences but its effects would be almost immediate (within years). This latter characteristic is commonly recognized as allowing us to "buy time" by rapid deployment in case of a climatic emergency.

The second category is carbon dioxide removal (CDR) and includes transport and storage of CO_2. Carbon dioxide removal is estimated to be more costly by comparison to SRM, and it has not yet been demonstrated on a large scale and it too could have negative local environmental consequences (particularly ocean fertilization; see below). In addition, the time to counteract greenhouse emissions by CDR would take much longer to operate – several decades as opposed to years for SRM. The fifth IPCC-AR5 report discusses both SRM (Boucher et al., 2013) and CDR (Ciais et al., 2013). Other options to mitigate climate change were discussed in Chapter 7 and together with climate engineering they are not mutually exclusive (e.g. Wigley, 2006).

The question as to whether it is even ethical to undertake climate engineering research was addressed by Robock (2012) who concluded

Climate Change in the Anthropocene.
DOI: https://doi.org/10.1016/B978-0-12-820308-8.00002-7

that modeling and laboratory experiments were ethical but that real life experiments were not, unless subject to global governance policies. The iron fertilization experiment in the 1990s in the Pacific Ocean (Watson et al., 1994) was performed in the open ocean, outside national exclusive economic zones, which makes the experiment legal, but according to Robock's (2012) criteria, unethical. Crutzen (2006) advised that real experiments should be done slowly and step by step and only after sufficient research has been done. Bodansky (1996) addressed the international legality of atmospheric climate engineering and concluded that the legal guidelines were weak and that it is in uncharted territory. He pointed out that the atmosphere above a state belongs to that state and that they can do as they wish, but also that atmospheric climate engineering affects all nations. The international rules that do exist, caution to do no harm and to avoid negative trans-boundary effects — but in atmospheric climate engineering there are likely to be winners and losers. International rules suggest that should a single or group of nations unilaterally decide to undertake a climate engineering project that the burden of proof would be on them to show that it was safe, otherwise it would not be allowed. Powerful or rogue nations might ignore this advice. Barrett (2008) suggests that because solar climate engineering is likely to be very low cost, on a per capita basis, it is an attractive option for nations to act unilaterally. Barrett (2008) also discusses how Nash equilibria among non-cooperating nations might control whether a nation unilaterally tries climate engineering or not, coming down to a cost-benefit analysis for each individual nation. The issue of sustainability is also relevant here given the long time-frames involved — will subsequent generations benefit or not from climate engineering projects? The goal of course is for a better long-term outcome, but the risks of harmful and catastrophic outcomes are unknown. Most workers agree that reducing greenhouse gas emissions is the preferred route to mitigation of warming (Crutzen, 2006; Caldeira et al., 2013; Chapter 7) but to date that route has been largely unsuccessful. This chapter addresses the scientific, rather than the political or social issues, of climate engineering under the two categories outlined above.

10.2 Solar radiation management

There are several variations of SRM but all attempt to reflect sunlight back into space and they include (Caldeira et al., 2013):
- Deployment of a large fleet of reflective satellites in space.
- Injecting aerosols into the stratosphere.

- Increasing the albedo of marine clouds.
- Making the oceans more reflective.
- Making plants on the continents more reflective.
- Increasing the albedo of the roof of buildings.

The first suggestion appears to fall into the realm of science fiction and the last proposal has a very small potential mitigation effect and then only in cities. Some of the other options are discussed below.

10.3 Aerosol injection into the stratosphere

The Earth absorbs 240 W of sunlight per meter squared (Chapter 3). Modeling indicates a doubling of CO_2 causes a radiative forcing of 4 W/m^2 (Matthews and Caldeira, 2007), so to offset this 4 W/m^2 requires reflection of 4/240 or 1.8 percent of incoming solar radiation. Many computer models do this simply by reduction of solar intensity, recognizing that solar intensity and CO_2 warming have different spatial and temporal effects (the sun does not shine at night and its intensity varies with latitude; these variations do not apply to CO_2 warming). Several different model results indicate changes in precipitation, the hydrological cycle, ocean current circulation, and variations between the poles and the tropics (Govindasamy et al., 2003; Lunt et al., 2008; Caldeira et al., 2013) but that reductions in solar radiation mentioned above can largely counteract the temperature effects of CO_2 warming. It is worth mentioning that solar climate engineering would not affect increasing ocean acidification, but lower temperatures would arrest or lessen sea level rise. Also, the magnitude of the solar flux reductions mentioned above ($\sim$ 2%) is five times lower than that of the longer term Milankovitch cycles (Chapter 4).

Crutzen (2006) discussed the injection of sulfate particles into the stratosphere and the possible consequences of such an albedo modification scheme. In the stratosphere, chemical and physical processes convert SO_2 into fine sulfate particles (0.1–1μm) and these particles reflect solar radiation leading to cooling. This was observed after the eruption of Mt. Pinatubo in 1991, which injected 10 Mt of sulfur into the stratosphere, causing cooling of 0.5°C the following year. The eruption also caused a global reduction in precipitation (Hegerl and Solomon (2009). The stratosphere is favored over the troposphere since the residence time for the particles there is much longer than in the troposphere, thereby requiring injection of less SO_2 for the same cooling effect. The delivery mechanisms include high altitude balloons lofted from ships or ocean islands or artillery cannons (Crutzen, 2006).

Workers at the Lawrence Livermore National Laboratory presented one of the first studies to use a climate model to evaluate an albedo modification scheme (Govindasamy et al., 2003). They adopted a global circulation model (Community Climate Model 3) which has a horizontal grid resolution of 2.8°, with 18 vertical levels. The model allows for ocean-sea ice interaction and the ocean has a prescribed heat flux (which limits flexibility to some extent). Three equilibrium experiments were performed:

- A preindustrial control experiment with 280 ppm CO_2 and a standard solar flux of 1367 W/m^2.
- A quadrupling of CO_2 (1120 ppm) with the same solar flux.
- A geoengineered quadruple CO_2 experiment with a solar flux reduced by 3.6%.

The models were run for 40 years and the averages for the last 15 years were reported. The geoengineered model counteracted the four times CO_2 concentration warming but with variations between the tropics and the equator and with a reduced seasonal variation. The hydrological cycle was not greatly affected but the authors note other models show significant hydrological effects (see below). The geoengineered model largely recovers sea ice thickness to the control experiment values. The response of the biosphere and the chemistry of the stratosphere to the climate engineering was not addressed.

Matthews and Caldeira (2007) modeled the transient response to increased CO_2 and solar climate engineering simultaneously over the period 1900 to 2100. They used an intermediate complexity model that took into account ocean circulation, heat uptake, sea-ice dynamics and atmosphere energy and moisture dynamics and terrestrial vegetation distribution. They also abruptly stopped climate engineering and modeled the transient effects of warming over time. They reduced incoming solar radiation by 1.8% in response to a doubling of CO_2. Radiative forcing was modeled as the natural logarithm of simulated CO_2 concentration relative to a reference preindustrial standard (280 ppm):

$$K_g S_{toa}(1 - \alpha) = F \ln (CO_2/280) \qquad (10.1)$$

where K_g is the climate engineering factor (0.018), S_{toa} is the solar flux at the top of the atmosphere, α is the Earth's new geoengineered albedo and F is a constant (5.35 W/m^2). The model brought the mean global temperature down to preindustrial levels within months. Increased CO_2 levels in the absence of surface warming, however, caused reduced evaporation and reduced precipitation over land—up to 1mm/yr. Such a large reduction could turn a moist region into desert within decades. One of the more startling results was that an abrupt stop in climate engineering (whether intentionally or accidently) resulted in a rapid temperature rebound twenty times (4°C/decade) the current heating rate due to the continuing accumulation of CO_2 while climate engineering was in operation.

With a more advanced global circulation model from the UK meteorological Office (HadCM31), Lunt et. al. (2008) undertook similar experiments to those of Matthews and Caldeira (2007) above. They report that in the geoengineered world the annual mean surface temperature is identical to the preindustrial model but with a reduced north-south temperature gradient, with cooler tropics. There is also a reduced intensity of the hydrological cycle, especially in the tropics. These observations are in agreement with Govindasamy et al. (2003). In addition, they find a significant reduction in Arctic sea ice and a decrease in temperature seasonality.

The Lawrence Livermore group also addressed the impact of solar climate engineering on the terrestrial biosphere in a separate study (Govindasamy et al., 2002). They performed four experiments using a coupled atmosphere-terrestrial biosphere model and mainly addressed vegetation distribution and biomass. The four (equilibrium) experiments were as follows:

- Control with 355 ppm CO_2 (circa 1992) with a solar flux of 1367 W/m^2
- Double CO_2 at 710 ppm, with the same solar flux
- Solar with CO_2 same as Control but solar flux reduced by 1.8%
- Geoengineered with double CO_2 and reduced solar flux by 1.8%

The distribution of vegetation was very similar in all four experiments, high biomass being controlled by warm moist climates and low biomass in deserts and high elevations and polar latitudes, as might be expected. Both climate and CO_2 controlled vegetation distribution equally whereas biomass was controlled largely by CO_2 concentration. Compared to the Control, Double CO_2 showed a temperature increase of 2.4°C, whereas the geoengineered experiment showed little change in temperature. Comparison of the geoengineered and the Double CO_2 experiments indicate CO_2 largely controls biosphere properties (biomass, net primary production, and soil respiration), which is known as the CO_2 fertilization effect. Biomass response was approximately double in the high CO_2 experiments compared to the Control experiment. The model did not address such factors as ecosystem services and goods, species abundance or migration, habitat loss, or ocean biosphere feedbacks. In short, the geoengineered world would be very different from today in terms of the terrestrial biosphere, but not in terms of temperature and with only slightly less precipitation. However, Matthews and Caldeira (2007) did see large reductions in precipitation over land in the tropics. Reduced incoming solar flux causes cooling with less evaporation and hence less precipitation (Hegerl and Solomon, 2009). Whether the opposing forces of higher CO_2 (higher biomass) together with increased drought would help or hinder agriculture is unclear and underscores the potential unforeseen consequences of climate engineering (i.e. there likely would be winners and losers).

The Mount Pinatubo eruption (June 1991) provided a natural experiment such that scientists could calibrate potential aerosol climate engineering experiments involving injection of aerosols into the stratosphere. This eruption had a cooling effect of about 0.5°C but did not destabilize the global climate. Wigley (2006), using the climate model MAGICC (which he and others designed), modeled the effects of three Pinatubo-type events by injecting 5 Mt of sulfur (50 percent of the Mt. Pinotubo event) into the stratosphere on an annual, two year, and four year schedule. These experiments had the effect of reducing temperatures by 0°C to −5°C over a hundred year period (the radiative forcings were up to negative $3W/m^2$). He coupled these experiments with three CO_2 concentration scenarios over the same time period. Long-term temperatures were confined to increases of 1°C to 2°C and the rate of sea level rise was also reduced. Wigley's 2006 study also confirmed the rapid temperature rebound effect on termination of solar climate engineering indentified in more detail by Matthews and Caldeira (2007). Combined solar climate engineering and CO_2 mitigation scenarios have the advantage of "buying time" without dramatic CO_2 reductions, and in addition, ocean acidity and sea level rise are both moderated.

It is difficult to compare the results of the above numerical models with each other because in most cases different types of experiments were performed. To address this problem the Climate Engineering Model Inter-comparison Project (GeoMip) involved 12 research groups using different global circulation models that performed the same experiment involving quadrupling the CO_2 concentration abruptly and reducing solar insolation (by about 4%) to bring temperatures down to preindustrial levels which all models succeeded in doing (Kravitz et al., 2013). The tropics were slightly cooler (−0.3°C) and the poles slightly warmer (0.8°C) and nearly all of the Arctic sea ice melting that would otherwise have occurred was prevented. Some tropical regions received less precipitation. Net primary productivity increased by 120% due to the CO_2 fertilization effect and reduced heat stress on plants. Fig. 10.1 shows a second group of idealized experiments performed

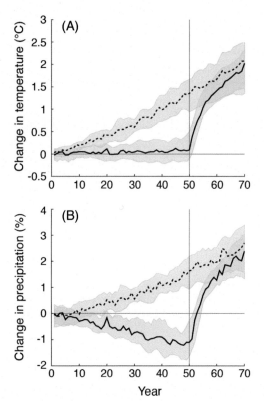

Figure 10.1 Idealized experiments (Climate Engineering Intercomparison Project): (A) Temperature. (B) Precipitation. Solid lines indicate solar radiation reduction required to match a 1% increase per year in CO_2. The dashed lines represent changes without solar radiation management (1% increase per year in CO_2). The experiments are terminated after 50 years and show a sharp temperature and precipitation rebound effect. Note reduction in global precipitation (part B) during the solar radiation experiment. *(Source: Boucher et al., 2013.)*

by the GeoMip group for globally averaged temperature (Fig. 10.1A) and precipitation (Fig. 10.1B). The climate engineering experiment is abruptly stopped after 50 years with the result that the temperature rapidly returns to what it would have been without climate engineering – a result we already saw above. Note that precipitation decreases globally by about 1 percent during the climate engineering experiment (Fig. 10.1).

The above workers all have concerns regarding changes in the biosphere, ocean acidification, and ozone destruction in the stratosphere. Although the above models were successful in reducing global mean annual temperature to preindustrial levels, the authors of these studies do not recommend solar

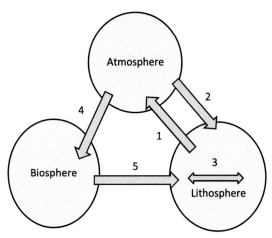

Figure 10.2 Summary of carbon dioxide reduction (CDR) methods involving three carbon reservoirs (atmosphere, lithosphere, biosphere): 1. Fossil fuel burning. 2. Direct CO_2 capture from the atmosphere; direct air capture with carbon capture and storage involves 2 and 3; also applies to land-based chemical weathering on a much longer time scale. 3. Carbon dioxide capture and storage within the lithosphere. 4. Afforestation. 5. Biomass energy; biomass energy with carbon capture and storage (BECCS) involves 4, 5, and 3.

climate engineering over reduction of greenhouse gas emissions because of associated negative effects, some possibly catastrophic. Climate engineering may, however, be necessary in case of climatic emergency (commonly taken as a greater than 2°C increase over preindustrial temperatures, as expressed in the Paris Accords; Chapter 7). At the time of writing (2021), the Paris Accords' goals of net zero emissions by 2050 are unlikely to be met globally.

10.4 Albedo enhancement of low-level marine clouds

Increasing the albedo of low-level marine clouds by injecting the clouds with saltwater particles that act as condensation nuclei is a second type of solar radiation management that increases the brightness of the clouds (Latham et al., 2008). Clouds reflect incoming UV radiation but also trap outgoing IR radiation to have an overall cooling effect of about 13 W/m^2. So, in principle, clouds could counteract the warming effect of CO_2 warming if their albedo could be increased. Global circulation models (GCM) provide a global perspective on cloud seeding experiments and Latham et al. (2008) used the HadGam numerical model (from the U.K. Meteorological Office). This model was used to estimate outgoing short-wave radiation flux at the top of the atmosphere among several other parameters. Global circulation models and simple calculations indicate that this idea can counteract the warming caused by doubling the CO_2 concentration by reducing radiative forcing by about 4 W/m^2, similar to the radiative forcing by the stratospheric injection scheme discussed above.

Important variables in calculating the change in cloud albedo (ΔA) are the fraction of the Earth covered by ocean ($f_1 = 0.7$), the fraction of the ocean covered by stratiform clouds ($f_2 = 0.25$) and the fraction of these clouds that are seeded ($f_3 = 0.1 - 1.0$). The relation between the change in radiative forcing (ΔF) and change in planetary albedo (A_p) is $\Delta F = -340\Delta A_p$ (where 340 W/m^2 is the solar irradiance at the Earth's surface; Chapter 2) and it follows the change in cloud albedo is:

$$\Delta A = \Delta A_p/(f_1 f_2 f_3) = \Delta F/60 f_3 \qquad (10.2)$$

For a decrease in radiative forcing of $\sim$ 4 W/m^2 the cloud albedo needs to change by about 0.066 (or 12 percent) assuming $f_3 = 1$ (all clouds seeded). Another important parameter is the ratio of existing cloud nuclei (N_o) and those added by seeding (N). The ratio is given by:

$$N/N_o = \exp(-\Delta F/4.5 f_3) \qquad (10.3)$$

To produce a forcing of -4W/m^2 (for $f_3 = 1$) the value of N/N_o is 2.4. If the ratio (N/N_o) is less than 0.3, Latham et al. (2008) estimate that a radiative forcing of -4 W/m^2 could not be realized (assuming f_3 between 0.2 and 1.0). The global average mean cooling effect at the top of the atmosphere over a five year period was -8W/m^2 twice that needed to counteract a doubling of CO_2 concentration. The results showed a strong seasonal effect and therefore also a large difference between hemispheres. These workers also estimate the rate of application of the seawater aerosol should be about 30 m^3/s, which is technically feasible (from ships, planes, or derricks). It has been suggested in the case of stationary derricks the power could be supplied by solar, wind, or wave energy.

The microphysics of aerosol-cloud interactions are complex and many aspects are still poorly understood. The fourth IPCC report discussed cloud-aerosol interactions and indicated that scientific understanding of these effects was "very low" (Table 7.10 in Denman et al., 2007). However, the fifth IPCC report devotes an entire chapter to clouds and aerosols and indicates important new advances in scientific understanding (Boucher et al., 2013). Some potential problems with the seeding cloud scheme and issues that need to be resolved are as follows. A global cloud seeding project would lead to global changes in ocean currents, temperature, and rainfall and the implications of these changes have yet to be fully evaluated. Since the proposed scheme is confined to the ocean, ocean-continent temperature contrasts would be altered and this effect needs to be further studied (Latham et al., 2008). Rainfall is likely to be decreased over the ocean and this may negatively affect some coastal communities.

10.5 Surface albedo enhancement

Since the land surface is only about one third of the planetary surface and approximately half is covered by clouds, approximately a 10% surface albedo increase would be needed to counteract the warming due to a doubling of CO_2 (Caldeira et al., 2013). Continent-based schemes of SRM would therefore seem to offer little promise and preliminary models bear this out. These methods include modification of cropland albedo (Ridgwell et al., 2009), modification of urban rooftops (Akbari and Rosenfeld, 2009), and increasing the albedo of the ocean by injecting small bubbles into the surface. Modification of the albedo of desert regions has also been studied by numerical models; the results suggest a profound alteration of the climate on a regional scale would result (Irving et al., 2011). None of these schemes seem to offer the potential to be implemented because either of their minor effects or their negative environmental toll.

10.6 Carbon dioxide removal

10.6.1 Introduction

The life time (or residence time) of fossil fuel CO_2 in the atmosphere was examined in an inter-comparison study of different numerical models, and it concluded that it was 200 to 2000 years, much longer than previously assumed (Archer et al., 2009). An important conclusion of such model studies is that if a pulse of CO_2 is injected into the atmosphere the climate will continue to warm for centuries into the future. Furthermore, such studies predict that if the climate is to be stabilized at a given temperature, then emissions must approach near zero (e.g., Matthews and Caldeira, 2007). This conclusion is clearly important in its own right, but also because it is easily understood by policy makers, and indeed several major emitter nations have pledged zero emissions by around midcentury, including China and the USA (under president Biden). Both the International Energy Agency

(IEA, 2020) and the Global Carbon Capture and Storage Institute have concluded that the Paris Accords' goal of keeping warming to less than 2°C cannot be met without removal of CO_2 by carbon capture and storage (CCS) from the atmosphere and not just by reducing emissions.

The long-term consequences of CO_2 removal were addressed by Cao and Caldeira (2010). One of the more startling conclusions of these authors is that in order to maintain low CO_2 concentrations and low temperatures that not only has the excess anthropogenic CO_2 in the atmosphere be removed, but in addition, all the anthropogenic CO_2 stored in ocean/land reservoirs after they outgas back to the atmosphere. If, for example, CO_2 concentrations were reduced by carbon capture to preindustrial levels, an additional amount equal to the captured amount would need to be removed over the subsequent 80 years. This means that CDR needs to be a long-term response. Although the simulations of Cao and Caldeira (2010) were extreme and idealized, the results nevertheless illustrate the response of the climate system to any removal of CO_2 at all scales and by all methods. Fig. 9.2 summarizes the exchange of carbon dioxide between three major reservoirs – the atmosphere, the lithosphere and the biosphere.

In addition to examining CCS, the sections below examine other means of removal of CO_2 from the atmosphere, including direct capture from air (DCCA), forestation/reforestation, ocean fertilization, accelerated weathering of rocks and biomass energy with carbon capture and storage (BECCS) (Table 10.1).

10.6.2 Carbon capture and storage

This topic was briefly introduced under the heading of mitigation in Chapter 7. The Global Carbon Capture and Storage Institute (GCCSI) is a CCS advocacy group and the following is a summary of part of their global status report for 2020. Two types of CCS facilities are recognized in the report: commercial facilities and pilot and demonstration facilities. In 2020 there were 65 commercial CCS facilities worldwide, 26 of which were operating, 3 were under construction and 21 were in early development; there were 34 pilot and demonstration facilities. The total amount of CO_2 captured and stored in 2020 was 40 Mt CO_2 (million metric tons) and most facilities captured between 0.2 and 5 Mt CO_2 per year. The IPCC Special Report on keeping warming to 1.5°C above preindustrial levels concluded that net-zero emissions must be achieved by 2050 if that temperature goal is to be met (see Chapter 8). By that time, 5.6 Gt CO_2

Table 10.1 Summary of selected carbon removal methods.

CDR method	Means of removal	Carbon storage/form	Time scale of storage	Potential C removed percentury
Afforestation/ reforestation	Biological	Land/organic	Decades to centuries	40–70 Gt C
BECCS	Biological	Geological/ inorganic	Permanent	125 Gt C
Ocean fertilization	Biological	Ocean/ inorganic	Centuries to millenia	280 Gt C
Enhanced weathering on land	Geochemical (carbonates/ silicates)	Ocean/soils/ inorganic	Centuries to millenia	100 Gt C
Direct air capture	Chemical	Geological/ inorganic	Permanent	No determined limit

Source with permission: Adapted from Table 6.15, Ciais et al., 2013. Gt C = 10^9 tonnes carbon.

per annum will need to be stored, more than a hundred fold increase over today.

Carbon capture and storage will be essential to reaching net-zero goals, especially in the power generation sector. The industries that are hardest to decarbonize are cement, iron, and steel, due to the nature of their industrial processes, and net-zero emissions maybe impossible in these industries. Aviation and shipping are also difficult to decarbonize. CCS power plants supply low-carbon electricity and help stabilize the grid and they complement renewable sources (solar, wind) by making the grid more reliable and resilient. Two technologies that potentially allow for negative emissions are biomass fuels combined with carbon capture and storage (BECCS) and direct air capture and carbon storage (DACCS), the latter of which is still in the experimental stage as of 2020.

Some examples of CCS facilities in 2020 are as follows:
- The Drax power station, UK, is being converted from a coal-fired power plant to a biomass fuel (wood pellets) with CCS (BECCS). Four Mt CO_2 are targeted to be captured per annum from one of its four units, the CO_2 to be sequestered in North Sea oil fields. All four units should be converted to BECCS by 2030.
- Enchant Energy is developing a CCS project that will capture 6 Mt CO_2 per annum at its coal-fired power plant in New Mexico, USA, and the CO_2 is to be used for enhanced oil recovery (EOR) in the Permian Basin.

- Santos Energy, Australia, has commenced a CCS project to capture 1.7 Mt CO_2 from its natural gas processing plant at Moomba. The CO_2 will be stored in nearby geologic formations. Santos estimates abatement costs of $22 (in US dollars) per tonne CO_2.
- The Alberta Carbon Trunk Line is the largest CO_2 capacity infrastructure in the world (14.6 Mt CO_2) and assumed operation in March 2020. The source capture facilities are an oil refinery and a fertilizer plant. The CO_2 will be used in EOR. The large capacity of the line allows for future expansion.

The irony of using captured CO_2 to extract more fossil fuels (EOR) in the examples above will not be lost on anyone (see Box 9.1). Because CCS infrastructure is very expensive, ranging from hundreds of millions to billions of dollars, policies are needed to incentivize investment such as tax credits, grant support, regulatory requirements, low cost storage, and low cost capture. Risks to investment in CCS include the interdependency of the infrastructure – if the source facility fails, the transport and storage components are left without support. The creation of hubs where storage and transport costs are shared would reduce costs and that is already underway in Europe. The low cost of CO_2 makes EOR an attractive option, so that a price needs to be placed on carbon to encourage reduction of emissions. It is estimated that in 2020 this price needs to be $40 - $80 per tonne CO_2 and increased to $50 - $100 per tonne by 2030. A time limit also needs to be placed on liability in case of leaks from geological storage; such long-term exposure to liability would discourage many investors (Box 10.1).

BOX 10.1 Carbon balance of enhanced oil recovery

The US Department of Energy (USDE, 2021) estimated, based on 114 enhanced oil recovery projects (EORs) in the U.S. since the year 2000, that the injection of 6×10^7 m^3 (2×10^9 ft^3) of CO_2 per day resulted in production of 2.8×10^5 barrels of oil per day. Assuming that the injected CO_2 is sequestered permanently, the question arises: what are the CO_2 emissions when this additional recovered oil is combusted and are they greater or less than the sequestered CO_2? In other words, is EOR a positive or a negative term on the balance sheet of emissions? The calculation here converts cubic feet of injected CO_2 and barrels of oil produced, when combusted, to kilograms of CO_2. The following conversions are used: one barrel (42 US gallons) of oil combusts to produce approximately 430 kg of CO_2 (Azzolina et al., 2016); one cubic foot of gas is equivalent to 28 liters at standard

(continued)

Box 10.1 Carbon balance of enhanced oil recovery—*cont'd*

temperature and pressure (STP) and 1 kilogram of injected CO_2 has a volume of 559 liters at STP, allowing cubic feet of gas to be converted to kilograms of CO_2.

The oil recovered (2.8×10^5 barrels per day) when combusted amounts to 120×10^6 kg or 120 tonnes CO_2 per day. The volume of injected CO_2 (2×10^9 cubic feet per day) corresponds to 10^8 kg or 100 tonnes CO_2:

100 tonnes CO_2 (injected) equivalent to 120 tonnes CO_2 per day after oil combustion.

Therefore, an additional 20 tonnes of CO_2 is produced per day, compared to the sequestered CO_2, when the oil recovered by EOR is combusted. This calculation excludes the emissions due to transport and refining of the oil. Although simplistic, the calculation here is based on a large number of EOR projects in the U.S. as the input, and the results of a much more comprehensive analysis by Azzolina et al. (2016) are broadly consistent with the conclusion here. Enhanced oil recovery is not a route to net zero emissions but it is far superior to combusting oil without CO_2 sequestration. Enhanced oil recovery will probably remain attractive on the basis of dollar costs since the recovered oil pays for the cost of the CO_2.

Global geological storage capacity is many times that required to reach net-zero emissions under any scenario. The oil industry has decades of knowledge on the geological and geophysical properties from seismic studies and drill holes from these reservoirs. These reservoirs, however, are not always located close to CO_2 producing facilities or in easy to reach locations. Much more common than oil/gas reservoirs are saline formations (98% of all storage capacity), but these formations are poorly characterized geologically and it is recommended that governments undertake studies of these vast potential storage reservoirs.

Potential storage reservoirs should be greater than 800 meters deep, to provide sufficient overburden pressure, and consist of both porous and impermeable formations. Knowledge of the rate of oil or gas production from a depleted reservoir allows the rate of CO_2 injection to be estimated and the volume of oil/gas produced allows the volume of the reservoir to be estimated with some confidence. That these reservoirs stored oil and gas over geological time scales permits the assumption that the sequestration of CO_2 will be permanent, but the probability of leakage cannot be assumed to be zero. As CO_2 is more dense than air, a large leak would displace the air at ground level and possibly suffocate nearby humans or livestock, as occurred in the Cameroon in 1986 (Chapter 7). The Gulf of Mexico well blowout of 2010 shows that wells can be improperly sealed even by experienced major oil companies (in this case British Petroleum). Offshore CO_2 storage reservoirs would present a lower risk, but at greater cost.

10.6.3 Afforestation/reforestation

Forests affect albedo, evapotranspiration and surface roughness all of which have climate consequences (Bonan, 2008). In terms of global carbon stock tropical forests show the largest ($\sim$1000 GtC), followed by temperate forests ($\sim$ 500 GtC) and by boreal forests ($\sim$250 GtC) and net primary production per year shows a similar pattern among these forest types. Tropical forests mitigate warming through evaporative cooling, whereas boreal forests have a positive radiative forcing due to the masking effect of snow albedo by the forest canopy which has a lower albedo. Boreal forests also store a large amount of carbon in soil, permafrost, and wetlands. Global warming causes these forests to expand into tundra and to die back along their southern borders. Much of the world's temperate forests have been cleared for agriculture and computer simulations indicate that trees lead to warming of the air temperature relative to crops. Loss of tropical forests in the 1990s amounted to about 150,000 km^2 per year (equivalent to the area of the state of Georgia, USA) and Amazonian forests were cleared at a rate of 25,000 km^2 per year at that time (Bonan, 2008). House et al. (2002) estimate that an ambitious afforestation program (reforestation of all previously cleared land) could decrease the atmospheric CO_2 concentration by the end of the century by 40 to 70 ppm, a relatively insignificant amount given current concentrations ($\sim$420 ppm) and an increase in emissions of 2.5 ppm/year.

10.6.4 Direct air capture

Keith et al. (2018) described a commercial pilot plant capable of capturing 1 million tonnes (Mt) of CO_2/year directly from air using two coupled chemical loops. The plant at full scale would be powered by renewable energy. They maintain that the cost would range from approximately \$100 to \$200 per tonne CO_2, considerably cheaper than previous DAC schemes. Most direct air capture proposals involve absorption into highly alkaline solutions or alternatively, adsorption onto solids, such as zeolites. The CO_2 could be stored in geological formations. An important point made by Cao and Caldeira (2010) is that if CO_2 and temperature are to be kept low, not only does anthropogenic CO_2 have to be removed from the atmosphere, but the CO_2 stored in the ocean and land must also be removed after degassing.

The New York Times (NYT, 2021) reported that some major corporations (e.g., Delta Airlines; Occidental Petroleum) are investing in DAC.

10.6.5 Biomass energy with carbon capture and storage

As mentioned earlier (Chapter 7), using biomass to generate electricity is considered carbon neutral and when combined with capture and storage of CO_2, it is carbon net negative – it removes CO_2 from the atmosphere. For it to be considered a type of climate engineering it must operate on a large or global scale so that a large source of biomass and therefore a large amount of land is required (e.g., temperate forests). The biomass also needs to be harvested at a sustainable rate – that is repeated harvesting of plants on the same land (Caldeira et al., 2013).

10.6.6 Land-based weathering

Although the kinetics of weathering reactions are very slow compared to the rate of CO_2 emissions, it has been proposed that accelerated weathering of carbonates or silicates could be used to sequester CO_2 from the atmosphere (Kelemen et al., 2011; Caldeira et al., 2013). Grinding the minerals down increases surface area and reaction rates and the fine grained minerals could be spread on agricultural land; alternatively, captured CO_2 from a point source (e.g. power plant) could be reacted in autoclaves at elevated temperature (Schuiling and Krijgsman, 2006). In the case of the dissolution of carbonate (limestone) the bicarbonate ion would be washed to the oceans:

$$CaCO_3 + H_2O + CO_2 \rightarrow Ca^{+2} + 2HCO_3^- \qquad (10.4)$$

In the case of silicate weathering, Olivine (Mg_2SiO_4) is one of the fastest reacting silicates and reacts with CO_2 to produce magnesite ($MgCO_3$):

$$4Mg_2SiO_4 + 8CO_2 \rightarrow 4SiO_2 \text{(quartz)} + 8MgCO_3 \qquad (10.5)$$

In these reactions the calcium and magnesium ions combine with the bicarbonate ion to form carbonates (limestone and dolomite) in the oceans, which are a major sink for CO_2. Olivine is a major component in rocks, such as peridotite and dunite, which are common in ultramafic intrusions, and also in formations called ophiolites which are slices of the Earth's upper mantle that were thrust onto the continents in mountain belts around the world. Under surface conditions the limestone reactions operate on a century time scale as can be seen in old graveyard limestone headstones where the names and dates are largely eroded. In addition, there exists a large amount of laboratory kinetic data for similar reactions.

10.6.7 Ocean fertilization

Photosynthesis involves the uptake of CO_2 (together with H_2O and sunlight) to produce of organic carbon molecules. Photosynthetic organisms such as phytoplankton in the shallow ocean produce organic molecules from inorganic carbon dissolved in the ocean and when they die some are carried to the deep ocean. The result is a reduction in the partial pressure of CO_2 in surface waters – in effect dissolved inorganic carbon is converted to organic carbon and transported to the deep ocean – this is called the biological pump in deep waters. The reduction in partial pressure causes a flux of CO_2 from the atmosphere to the shallow ocean. The concept of ocean fertilization is an attempt to draw down CO_2 from the atmosphere by enhancing the productivity of phytoplankton by adding nutrients to the ocean – the important nutrients being iron, nitrate and phosphate (Caldeira et al., 2013).

In an experiment carried out in the equatorial Pacific Ocean south of the Galapagos islands, Watson et al. (1994) introduced 437 kg of iron to an 8 km by 8 km patch of ocean and monitored the partial pressure of CO_2 over several days. Compared to laboratory experiments where introduction of iron resulted in total consumption of the other nutrients, phosphate and nitrate, the ocean experiment only produced a 10 percent CO_2 drawdown, so that the experiment was not effective in sequestering CO_2. Subsequently, Buesseler et al. (2008) presented a list of research questions that need to be answered before ocean fertilization should be tried on a larger scale. They point out that biochemical and ecological impacts are not well understood, especially on a longer time scale, with possible unforeseen consequences.

10.7 Discussion

Computer models on solar reduction management (SRM) summarized above agree that injection of particles into the stratosphere or the seeding of marine clouds to enhance albedo could keep temperatures in check by reducing incoming solar radiation and the effects could be seen in a matter of months or years. Potential environmental effects include changes in the hydrological cycle, ocean currents, seasonal variation, decreased rainfall and ocean acidification. The CO_2 fertilization effect on land would enhance biomass production. Should such a SRM system fail for any reason, a rebound effect would result in catastrophic warming. A global attempt at SRM would require international policies and governance, which do not yet exist.

Carbon dioxide removal (CDR) and storage proposals would need to operate on a longer time scale than SRM (i.e., centuries) to be effective in CO_2 mitigation and have not yet been shown to operate on a large scale. In addition the cost may be prohibitive. Aforestation/reforestation requires large land areas as does biomass with carbon capture and storage (BECCS). Direct air capture (DAC) is still in the experimental stage. Ocean fertilization has unknown potentially hazardous consequences for ocean ecosystems. The concluding statement of Caldeira et al. (2013) in their review of climate engineering namely that neither SRM or CDR can provide the certain reduction in risk that is offered by reduction in greenhouse gas emissions. That conclusion still stands.

References

Akbari, H., Menon, S., Rosenfeld, A., 2009. Global cooling: increasing world-wide urban albedos to offset CO_2. Clim. Change 94, 275–286.

Archer, D., Eby, M., Ridgewell, A., Cao, L., Mikolajewicz, U., et al., 2009. Atmospheric lifetime of fossil fuel carbon dioxide. Ann. Rev. Earth Planet. Sci. 37, 117–134.

Azzolina, N.A., Peck, W.D., Hamling, J.A., Gorecki, C.D., Ayash, S.C., et al., 2016. How green is my oil? A detailed look at greenhouse gas accounting for CO_2-enhanced oil recovery (CO_2-EOR) sites. Int. J. Greenh. Gas Control 51, 369–379.

Barrett, S., 2008. The incredible economics of climate engineering. Environ. Res. Econ 39, 45–54.

Bodansky, D., 1996. May we engineer the climate? Clim. Change 33, 309–321.

Bonan, G.B., 2008. Forests and climate change: forcings, and the climate benefits of forests. Science 320, 1444–1449.

Boucher, O., Randall, D., Artaxo, P., Bretherton, C., Feingold, G., et al., 2013. Clouds and aerosols *in* climate change: the physical basis. In: Stocker, T.F., Qin, D., Plattner, G.-K., Tignor, M., A, et al. (Eds.), Contribution of Working Group 1 to the Fifth Assessment Report of the IPCC. Cambridge University Press, Cambridge, UK.

Buesseler, K.O., Doney, S.C., Karl, D.M, Boyd, P.W., Caldeira, K., et al., 2008. Ocean iron fertilization – moving forward in a sea of uncertainty. Science 319, 162.

Caldeira, K., Govindasamy, B., Cao, L., 2013. The science of climate engineering. Ann. Rev. Earth Planet. Sci. 41, 231–256.

Cao, L., Caldeira, K., 2010. Atmospheric carbon dioxide removal: long-term consequences and commitment. Environ. Res. Lett. 5, 024011.

Ciais, P., Sabine, C., Bala, G., Bopp, L., Brovkin, V., et al., 2013. Carbon and other biogeochemical cycles *in* climate change: the physical basis. In: Stocker, T.F., Qin, D., Plattner, G.-K., Tignor, M.A, et al. (Eds.), Contribution of Working Group 1 to the Fifth Assessment Report of the IPCC. Cambridge University Press, Cambridge, UK.

Crutzen, P.J., 2006. Albedo enhancement by stratospheric sulfur injections: a contribution to resolve a policy dilemma? Clim. Change 77, 211–219.

Denman, K.L., Brasseur, G., Chidthaisong, A., Ciais, P., Cox, R.E., et al. 2007. Couplings between changes in the Climate System and Biogeochemistry in Climate Change 2007. The Physical Basis. In: Solomon, S., Qin, D., Manning, C.Z., Marquis, M., et al. (Eds.). Contribution of Working Group 1 to the Fourth Assessment Report of the IPCC.

Govindasamy, B., Caldeira, K., Duffy, P.B., 2003. Climate engineering Earth's radiation balance to mitigate climate change from quadrupling of CO_2. Glob. Planet. Change 37, 157–168.

Govindasamy, B., Thompson, S., Duffy, P.B., Caldeira, K., Delire, C., 2002. Impact of climate engineering schemes on the terrestrial biosphere. Geophys. Res. Lett. 29, 2061–2064.

Hegerl, G.C., Solomon, S., 2009. Risks of climate engineering. Science 325, 955–956.

House, J., Prentice, I., Le Quéré, C., 2002. Maximum impacts of future reforestation or deforestation on atmospheric CO_2. Glob. Clim. Biol. 8, 1047–1052.

IEA, 2020. International Energy Agency. https://www.iea.org (Accessed Mar 2021).

Irving, P.J., Ridgewell, A., Lunt, D.J., 2011. Climatic effects of surface albedo climate engineering. J. Geophys. Res. 116, D24112.

Keith, D.W, Holmes, G., St. Angelo, D., Heidel, K., 2018. A process of capturing CO_2 from the atmosphere. Joule 2, 1573–1594.

Kelemen, P.B., Matter, J., Streit, E., Rudge, J.F., Curry, W.B., Blusztajn, J., 2011. Rates and mechanisms of mineral carbonation in peridotite: natural processes and recipes for enhanced in situ CO_2 capture and storage. Ann. Rev. Earth Planet. Sci. 39, 545–576.

Kravitz, B., Caldeira, K., Boucher, O., Robock, A., Rasch, P.J., et al., 2013. Climate model response from Climate engineering Model Intercomparison Project (GeoMiP). J. Geophys. Res. Atmos. 118, 8320–8332.

Latham, J., Rasch, P., Chen, C.C., Kettles, L., Gadian, A., 2008. Global temperature stabilization via controlled albedo enhancement of low-level maritime clouds. Phil. Trans. R. Soc. London, A 366, 3969–3987.

Lunt, D.J., Ridgewell, A., Valdes, P.J., Seale, A., 2008. Sunshade World": a fully coupled GCM evaluation of the climatic impacts of climate engineering. Geophys. Res. Lett. 35, L12710.

Marchetti, C., 1977. On climate engineering and the CO_2 problem. Clim. Change 1, 59–69.

Matthews, H.D., Caldeira, K., 2007. Transient climate-carbon simulations of planetary climate engineering. Proc. Nat. Acad. Sci. USA 104, 9949–9954.

NYT, 2021. Businesses explore practical ways to pull greenhouse gases from the air. N.Y. Times Jan. 19, 2021.

Ridgewell, A., Singarayer, J.S., Hetherington, A.M., Valdes, P.J., 2009. Tackling regional climate change by leaf albedo bio-climate engineering. Current Biol 19, 146–150.

Robock, A., 2012. Is climate engineering ethical?, 2012. Sicherh. Frieden 30, 226–229.

Schuiling, R.D., Krijgsman, P., 2006. Enhanced weathering: an effective and cheap tool to sequester CO_2. Clim. Change 74, 349–354.

USDE, 2021. https://www.energy.gov/nepa/downloads/cx-020132-developing-co2-eor-and-associated-storage-within-residual-oil-zone-fairways. Accessed Jan. 2022.

Watson, A.J., Law, C.S., Van Scoy, K.A., Millero, F.J., Yao, W., et al., 1994. Minimal effect of iron fertilization on sea-surface carbon dioxide concentrations. Nature 371, 143–145.

Wigley, T.L.M., 2006. A combined mitigation/climate engineering approach to climate stabilization. Science 314, 452–454.

Index

Page numbers followed by "*f*" and "*t*" indicate, figures and tables respectively.

A

Adaptation, 105, 106, 108
  economics, 115
  optimal, 116
  options, 108, 110*t*
Adaptive strategies, 112
Aerosols, 29
  cloud interaction, 29
  injection, 169
  interactions, 29
  light–absorbing, 29
Agriculture and rural communities, 91
Albedo enhancement, 174
Albedo feedbacks, 34
Albedo values, 34, 43
Alberta Carbon Trunk Line, 179
Amazonian forests, 181
Anthropogenic climate drivers, 24
Anthropogenic greenhouse gas, 22, 29
Anthropogenic sources, 27
Arctic sea, 100
Arctic sea ice, 152
Argo floats, 51
Array for Realtime Geostrophic
  Oceanography, 48
Atmosphere-Ocean-General Circulation
  Models (AOGCMs), 41, 52
Atmosphere-ocean global climate
  models, 33
Atmosphere-ocean interaction, 101
Atmospheric Chemistry Climate Model
  Intercomparison Project
  (ACCMIP), 58
Atmospheric warming, 34

B

Biomass
  energy, 182
  response, 171
  storage, 184
Biome shifts, 150

Black body radiators, 22*t*
Bølling warming, 63
Boreal forests, 180

C

Carbon
  balance of enhanced oil recovery, 179
    180–179 180
  capture, 132
  cycle, 41
  dioxide, 25
  emissions, 124
  intensity of energy, 129
  prices, 161
  removal methods, 178*t*
  storage, 132
Carbon dioxide reduction methods, 174*f*
Carbon dioxide removal, 167, 176, 184
Clausius-Claperyon relation, 29
Climate
  engineering, 167
  impacts, 81
  models, 41
  processes, 41
  related processes, 34*t*
  response, 30
Climate Engineering Model
  Intercomparison Project, 172
Climate Model Intercomparison Project
  phase 5 (CMIP5), 57
Cloud-aerosol interactions, 33
Cloud-radiation interactions, 33
Coastal city flooding, 93
Computer climate models, 41
Computer models, 67
  project, 36
Continental ice sheets, 34
Coupled Model Intercomparison Project,
  45, 116
Cryosphere, 10
  ocean system, 34

D

Deoxygenation, 88
Deteriorating water infrastructure, 84
Direct air capture (DAC), 181, 184

E

Earth
    climate system, 3
    orbital changes, 66f
    sciences, 3
    surface radiation budget, 19
    system models, 41, 43, 58
    system models of intermediate
        complexity, 43
Eccentricity, 65
Ecosystems
    based adaptation, 114
    disruption, 150
    food webs, 151
Emissions, 133
    drivers, 127
Energy, 109
    demand, 119
    intensive processes, 134
Ensembles, 45
    multimodel, 45
    perturbed parameter, 45
Equilibrium climate sensitivity (ECS), 30
    values, 30
ERF calculations, 23
Evaporation, 36
Exposure, 105
External variability, 54
Extract more fossil fuels (EOR), 179

F

Fast exchange domain, 34
Feedbacks
    long-term, 34
    quantifying, 31
    short-term, 32
Fossil fuel development, 139
Freshwater, 109

G

Global
    atmospheric water, 6

carbon project, 52
deforestation, 14
energy budget, 20f
energy demand, 140
geological storage capacity, 180
mean sea level, 149
radiation budget, 19
warming, 151
warming potential, 24, 124
Gravity Recovery and Climate Experiment
    (GRACE), 13
Greenhouse effect, 19
Greenhouse gas, 25, 167

H

Hadley cell circulation, 8
Halocarbons, 28
High Plains Ogallala aquifer, 82
Holocene epoch, 73
Hydroelectric production schedules, 81
Hydrogen-based fuels, 164

I

Increased morbidity and mortality, 153
Intended Nationally Determined
    Contributions, 124
Intergovernmental Panel on Climate
    Change, 123
International Energy Agency (IEA), 123, 157
International Meteorological
    Organization, 4

L

Land
    based weathering, 182
    ocean cryosphere components, 12t
    processes, 14
    surface temperature, 5
    use decisions, 85
Laurentian ice sheet, 34
Lawrence Livermore National Laboratory,
    170
Light-absorbing aerosols, 29
Long-term feedbacks, 34
Long-term projections, 57, 60
LST datasets, 6

M

Marine
 ecosystems, 35
 organisms, 151
 plants, 35
Methane, 27
 emissions, 25
Milankovitch cycles, 65
Mitigation, 123
Montreal Protocol, 28
Mount Pinatubo eruption, 172
Multimodel ensembles, 45

N

National Climate Assessment Report, 19
Nationally Determined Contributions
  (NDCs), 159
Natural gas based ammonia
  production, 136
Near-term
 climate projections, 54
 projections, 54
 warming projections, 123
Net primary production, 35, 172
Net-zero emissions, 160
Nitrous oxide, 26
 values, 25
Nuclear energy, 132

O

Oceans, 48, 151
 ecosystems, 87
 warming, 35
Orbital forcing, 65
Organized intercomparison projects, 45
Ozone, 22, 28
 stratospheric, 28
 tropospheric, 28

P

Paleoclimates, 51, 63
Paleoclimatic records, 65
Parameterization, 42
Paris Accords, 130
 agreement, 125
 and representative climate pathways, 52,
  53

Paris Agreements, 53, 145, 159
pathways, 53
Permafrost, 38
Perturbed parameter ensembles, 45
Phenology, 150
Photochemical destruction, 26
Photosynthesis, 183
Photosynthetic organisms, 183
Polar amplification, 12
Potential storage reservoirs, 180
Pre-satellite reconstructions, 67
Projections of energy prices, 161

Q

Quantifying feedbacks, 31

R

Radiative forcing (RF), 22, 24, 170
Reference gas, 24
Reforestation, 85
Representative climate pathways (RCPs),
  14, 51, 52, 54
Representative Concentration Pathways,
  138
Representative pathway scenarios (RCPs),
  55
Residential energy, 137
Risk, 105

S

Sea surface temperature, 6
Shared socioeconomic pathways, 116, 137
Shared socioeconomic pathways (SSPs), 54,
  116, 120
Short-term feedbacks, 32
Shortwave radiation, 3
Socioeconomic pathways, 120
Soil erosion, 92
Soil protective measures, 92
Solar radiation management (SRM), 167
Solar reduction management, 183
Standardization procedure, 5
Stated Policy Scenarios, 158
State-of-the-art climate models, 42
Stefan-Boltzmann
 constant, 20
 law, 20, 22

Stratospheric ozone, 28
Stratospheric water vapor, 29
Surface albedo enhancement, 176

T

Terrestrial permafrost, 13
Top panel, 149f
Total energy supply, 162
Transient climate response (TCR), 30
Tropospheric ozone, 28

U

Urbanization, 92, 112
Urban social inequality, 113
U.S. National Climate Assessment, 124

V

Volcanic
    eruptions, 24
    forcing, 67
    injections, 67

W

Warmer ocean temperatures, 96
Warmer surface water, 88
Water vapor, 29, 33
Weather forecasting, 44
Wien's law, 20
World Meteorological Organization
    (WMO), 4
World Weather Records (WWR), 4

Printed in the United States
by Baker & Taylor Publisher Services